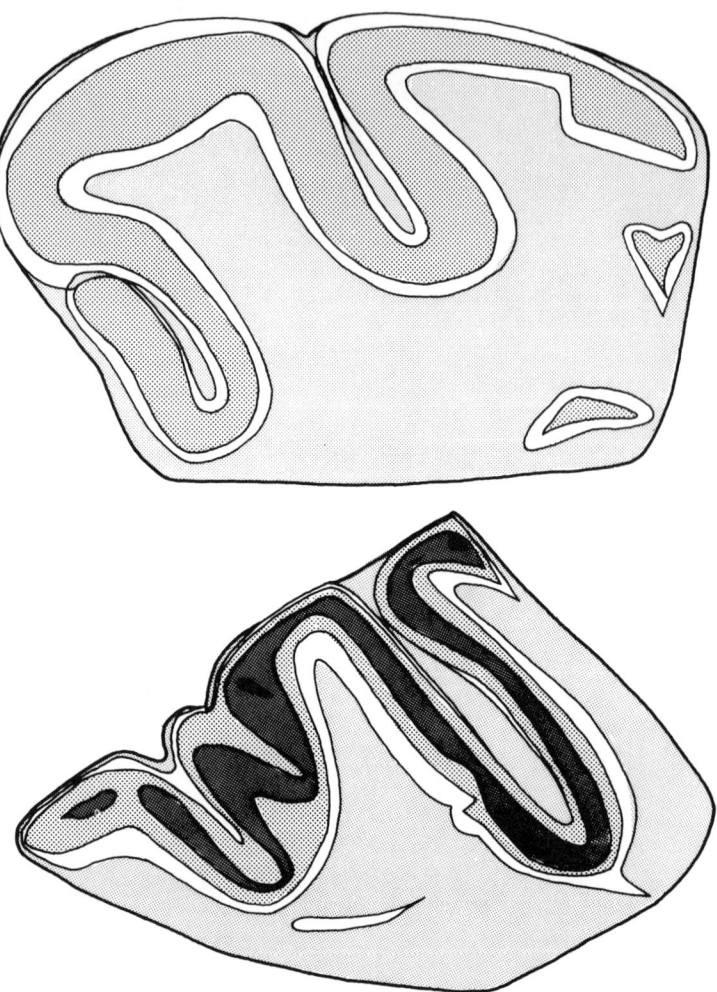

Frontis illustration: Density of serotonin receptors in brains of suicide (*bottom*) and control (*top*). Images are based on subtracted autoradiograms representing the total binding sites specific for serotonin 5-HT$_2$ in the prefrontal cortex (Brodmann area 9) of the human brain. The images are from a matched pair of subjects, with brains shown in the same calibration. The distribution of 5-HT$_2$ receptors is greater in the intermediate layers in both brains, with the increased binding at all layers of the suicide's brain; greater binding is associated with lower levels of serotonin. Reproduced from *Science,* vol. 254 (1991), p. 1451. Copyright 1991 by the AAAS. For a full report, see Victoria Arango et al., "Autoradiographic Demonstration of Increase Serotonin 5-HT$_2$ and β-Adrenergic Receptor Binding Sites in the Brain of Suicide Victims," *Archives of General Psychiatry,* vol. 47 (1990), pp. 1038–47.

The Neurotransmitter Revolution

Serotonin, Social Behavior, and the Law

Edited by
Roger D. Masters and Michael T. McGuire

With a Foreword by Margaret Gruter

Southern
Illinois
University
Press

Carbondale
and
Edwardsville

Library of Congress Cataloging-in-Publication Data

The Neurotransmitter revolution : serotonin, social
    behavior, and the law / edited by Roger D. Masters and
    Michael T. McGuire; with a foreword by Margaret
    Gruter.
        p.   cm.
    Includes bibliographical references and index.
    1. Forensic neuropsychology. 2. Neurotransmitters.
3. Serotonin. 4. Neurochemistry. I. Masters, Roger D.
II. McGuire, Michael T., 1929– .
[DNLM: 1. Behavior—drug effects—congresses. 2.
Legislation, Medical—United States—congresses. 3.
Neurochemistry—congresses. 4. Neuroregulators—
physiology—congresses. 5. Serotonin—physiology—
congresses. QV 33 AA1 N4]
RA1147.5.N5   1994
614'.1—dc20
DNLM/DLC
for Library of Congress   92-49968
ISBN 0-8093-1792-3   CIP
ISBN 0-8093-1801-6 (pbk.)

Chapter 4, "The Basics of Serotonin Neurochemistry," by Arthur Yuwiler, Gary L. Brammer, and K. C.
    Yuwiler, and chapter 6, "Serotonin and Violent Behavior," by Markku Linnoila et al., are in the public
    domain.

Contents

Figures

Tables

Foreword

The publication of *The Neurotransmitter Revolution* is welcome for many reasons. This book brings into print the fruits of a conference, "Serotonin, Social Behavior and the Law," which the Gruter Institute for Law and Behavioral Research organized and cosponsored in conjunction with Dartmouth College's Rockefeller Center for the Social Sciences. The meeting took place at Dartmouth on November 4–5, 1988. Since that time, it has become apparent that its contents should be available to lawyers, students, and scholars in many disciplines, as well as to a broader public of informed citizens.

Why should research in the biological sciences be studied by lawyers, economists, political scientists, and other students of human behavior? Over a decade ago, I formed the Gruter Institute to provide an answer to this question. The conferences and publications of the institute have been based on four ideas: first, the realization that biology is making rapid advances in its understanding of human behavior; second, that this understanding has significant implications for law; third, that biological understanding is advancing at a more rapid rate than it is being incorporated into law; and finally, that in many instances there are no well-established processes by which such understanding can be readily and predictably incorporated into law.

The Dartmouth conference thus continues in a series of meetings that have led to important books, beginning with the First Monterey Dunes Conference in 1981, published as *Law, Biology and Culture* two years later under the coeditorship of Paul Bohannan and myself. Subsequent conferences have led to the publication of two additional volumes, *Ostracism: A Social and Biological Phenomenon* (Elsevier, 1986) and *The Sense of Justice* (Sage Publications, in press), both of which I coedited with Roger Masters. Along with my own *Law and the Mind* (Sage Publications, 1991), these activities were intended to establish what I have long called "the ethology of law" as a way of understanding the relationship between the biological sciences, human behavior, and the law.

It sometimes comes as a surprise to biologists that the term "ethology" is often unknown when mentioned among well-educated people. Unfortunately, this is the case. Except for the most recent editions, most dictionaries define ethology as "the science of character," based on the use of the term dating from the 19th century. The 1985 Webster's gives as a second meaning "the scientific and objective study of animal behavior, esp. under natural conditions." I believe that Niko Tinbergen's definition of ethology as "the biological study of behavior" is adequate.

To speak of "the ethology of law" seems daring. To introduce a course entitled "The Ethology of Law" in a leading law school, such as the one taught by E. Donald Elliott and Roger D. Masters at Yale in 1988, seems even more adventurous. But the favorable response to such initiatives is very gratifying to those who have joined with me in the Gruter Institute, seeking to use ethology as Tinbergen defined it to help understand the conflicts in today's complex society that arise due to rapid new developments in the natural sciences.

The idea of an ethology of law goes back to my work in the early 1970s, when I came across Konrad Lorenz's article entitled "The Weltbild of the Ethologist." It has come a long way since. What does the term "ethology of law" mean? Probably it will mean different things to different people. Let me give you my version.

I start from the assumption that law is one of the most wondrous creations of the human brain. Evolving over millions of years from earlier evolutionary forms, the human brain has served the needs of creatures living over an extraordinary span of time. It is now like a composite—the "triune brain" as Paul MacLean calls it. This human brain has designed rules and has shaped human behavior as long as humans have existed. Its rule making was adaptive to the changing environment, just as the behavior of our ancestors had to be.

Our very existence today testifies to the fact that the evolving brain has mastered the task of changing or adapting rules to accommodate the demands of the ever-changing environment, keeping a balance between change and continuity of the legal structure that the group-living humans need to survive. Continuity and change in man-made law has enabled humans not only to survive but to dominate all other living species on this planet.

Based on this evolutionary view, law as a creation of the living brain becomes the "living law," an intricate mechanism, a part of our biological endowment. The term "living law," as it is used here, comprises the legal system, specific laws and their enforcement, and above all, legal behavior. Law over time has acquired its own life, and like a living organism it grows in itself, responding to the various ideological or religious beliefs with their dictates from the limbic system and the rational considerations originating in the neocortex of the human—triune—brain. While itself a biological creation, its functions at times may deviate from those of biologically based behaviors. This is most likely to happen when law attempts to develop new rules that are designed to restore a balance between rational thought processes and emotionally based moral expectations. To be aware of this underlying biological connection is critical for understanding the ethological view of the legal system.

Now back to the concept of ethology. Ethology is a branch of biology. It is the observational arm of biology and combines field and laboratory experiments with observations of animal species in nature. Through these methods, ethologists seek to identify rules of behavior vital to the survival of species in their environments.

Ethology concentrates on four main fields of study: causation (causal factors of motivation), development (interaction of genes and environment), evolution (linking specific research with other branches of biology, such as ecology or sociobiology), and function (explaining how various actions have come to be favored by natural selection and why people under particular circumstances tend to behave in predictable ways).

It is in the studies dealing with function that legal scholars may most easily find information relevant to legal problems of today. We speak of functions of behavior and also of functions of specific laws or legislation. The usages of the term "function" in law and biology may differ, yet by exploring the interrelationship of their meanings we may gain insights into the effectiveness of law in general and of specific legislation or laws. Such insights will be important if we postulate that *a given law will be more effective if the function of the law complements the function of the behavior that is regulated or channeled by the law*.

With this in mind, I propose that legal research and practice will keep pace more effectively with changes in human society if findings from the basic sciences are known, understood, and incorporated into legal thinking and practice. The goal is to connect the "brain child" law to its origins in human behavior, using insights gained in biology to challenge some new or old rules, their functions, and their effectiveness.

This volume is an outstanding example of how the ethology of law can contribute to a better understanding of the legal and social implications of the discoveries of contemporary biology. As with all collaborative efforts, it owes much to the help of a number of dedicated people. I am particularly thankful to the Rockefeller Center for the Social Sciences at Dartmouth College: its director, Richard Winters, and its staff, led by Roxanne Waldner, graciously and effectively supported our conference both financially and administratively. Roger D. Masters organized the meeting and has seen the present volume to completion despite a number of frustrating obstacles. Michael T. McGuire has again provided his intelligence and expertise in bringing the ideas of biologists together with the approaches of lawyers and social scientists. And above all, I thank Gerti Dieker, whose continued energies and efficiency have played such an essential role in the life of the Gruter Institute.

Dr. Margaret Gruter, President
The Gruter Institute for Law and Behavioral Research

Editors' Preface

Extraordinary advances in research on the neurochemistry of behavior have created an urgent problem: how can scientific discoveries about the workings of the human brain best be integrated into our legal system? Specialists are increasingly discovering how neurotransmitters and neuroanatomical structures influence human behavior. Does this mean that knowledge of physical or chemical causes will replace the concept of moral and legal responsibility? How should our society respond to the discovery that an individual's personality or actions were "caused" by specific events or anomalies in the brain?

This volume seeks to illustrate such issues by focusing on one of the many neurotransmitters whose effects on human behavior are being uncovered by the latest research. To symbolize the radical transformations in our knowledge about human behavior and the changes in our legal system that seem destined to result, we have called the volume *The Neurotransmitter Revolution: Serotonin, Social Behavior, and the Law*. This title seems more than justified, for the research we discuss goes to the very roots of deeply held beliefs and theories about human nature.

The specific chemical that is the focus of this book is serotonin. The implications of these essays, however, are clearly not limited to behavioral effects attributed to that one neurotransmitter and to the policy proposals that might result from such discoveries. This volume is intended to be a model of the kind of analysis that is necessary if our society and our legal system are to adapt to revolutionary changes in knowledge about the human brain.

While the diverse and powerful behavioral effects of serotonin (5-hydroxytryptamine) are increasingly understood, similar findings are emerging for many other neurochemicals, including dopamine, norepinephrine, GABA (gamma-aminobutyric acid), testosterone, estrogen, the enkephalins, and a long list of amino acids whose interactions are as bewildering as they are complex. But complexity is not the same thing as mystery. Neuroscience is discovering how the brain works—and chemistry is becoming one of the keys to these discoveries.

These scientific findings are quite different from what might have been expected even a decade ago. The neurochemistry of behavior is not the same thing as genetic determinism. On the contrary, neurotransmitters like serotonin vary from one individual to another for many reasons, including the individual's life experience, social status, and diet. Genes may influence neurochemistry. So do behavior, culture, and the social environment. Nature and nurture interact—but these interactions can be understood from a rigorously scientific point of view.

The specific discoveries concerning the connections between serotonin and such behaviors as suicide, seasonal depression, alcoholism, impulsive homicide and arson, and social dominance are thus important but not conclusive. More essential is the realization that we will need to confront a cascade of such findings in coming years. And such issues will require new interactions between specialists in the disciplines represented by the authors of the following chapters.

Social scientists and lawyers will increasingly need to consider the way scientific evidence can uncover chemical influences on behavior. Conversely, biologists and physicians will need to consider the legal and moral implications of their own research. Hasty conclusions by uninformed lawyers, journalists, or politicians could do unforeseen damage to the fundamental principles of a democratic society. It is thus time to attend to the "neurotransmitter revolution" before its effects outstrip our ability to comprehend them.

It is by no means easy to translate new scientific information about the causes of human behavior into changes in law, public policy, or social norms. In many respects, the law evolves solutions to new problems through time. The conscious intentions of lawyers, legislators, and scholars often do not determine the outcome. In fact, premature attempts to shape law or custom may have effects that are exactly the opposite of those that were intended.

On the other hand, the initial response to a new scientific discovery may make a significant difference over the long run. In disciplines as diverse as physics, meteorology, paleontology, history, military affairs, and economics, scholars have demonstrated the enormous effects of apparently miniscule differences at critical moments. Given the disquieting nature of the discovery that much human behavior is influenced by neurotransmitters like serotonin, those with a sophisticated understanding of the law and human social behavior would do well to consider how best to apply the findings described in this volume.

This volume is thus not devoted to a final set of policy recommendations. The specific conclusions suggested by one or another author are less important than the illustration of an informed dialogue between specialists in neuroscience, ethology, law, political science, and psychology.

Controversy is of the essence in issues of public policy. But it can hardly be controversial to hope that the introduction of scientific knowledge concerning the neurochemistry of behavior will not undermine the principles of a free society under law. Do our basic notions of legal responsibility need to be changed? How can the discovery of the physical mechanisms shaping human behavior be integrated with the legal and moral traditions that were based on the notions of freedom and free will?

What are the best policies to adopt in response to radical new discoveries in neuro-science?

Such questions are far from trivial. They need to be the focus of the attention of the best minds in our universities, our courts, and our political institutions. Should the next generation fail to consider new scientific findings with an open mind, it is hard to underestimate the dangers for behavior control and legal excess in the hands of bureaucrats or demagogues. If a responsible legal and political order is to survive the explosion of scientific knowledge, an interdisciplinary dialogue is needed. *The Neurotransmitter Revolution: Serotonin, Social Behavior, and the Law* is intended to contribute to this end.

A NOTE ON THE REFERENCES

The conventions for citing references differ in the varied academic disciplines repre-sented by the authors in this volume. Although it is usual to impose uniformity in style throughout a volume of collected essays, after careful consideration we have chosen to preserve the customary reference style used by each author. The reader deserves an explanation of this decision.

Our goal is a dialogue between scholars in the law or the social sciences, researchers in the life sciences, and the informed citizen. To this end, readers will need to assimilate research in disciplines with which they are not familiar. Because each field has its conventions, it will be important for the biologist to understand the referencing procedures in the legal profession, just as those in the law will need to be familiar with the forms of citation used by scientists.

Often, there are subtle reasons that might explain these disciplinary practices. Legal scholars, like those in the humanities, use many footnotes in which sources—and especially judicial decisions—are listed either at the bottom of the page or at the end of an article; footnotes with legal citations often include a sentence or two, with the leading point or precedent stated succinctly. Biologists and many social scientists now use a different system, in which parenthetical citations in the text refer to an alphabetical list of references at the end of the chapter. Physicists and researchers in some other experimental fields use a system with numbered references in the order of citation at the end of the article. Some disciplines, like history and political science, employ one or another of these approaches depending on subfield or individual preference.

The system used in legal scholarship and in the humanities is particularly relevant in an area where each citation refers to a substantive point or precedent—and where the context of the citation is as important as the reference. In contrast, the alphabetical listing of sources used by biologists and psychologists seems more functional when the contents of a citation are evident in the title, when a single reference may be cited at several points in a scholarly work, and when the relevance of a citation does not depend on the context. In disciplines where one point of a research paper needs to have citations to a large number of references, the numerical listing in order of citation (as in physics) may be most practical. Putting references at the end is also convenient in disciplines where readers may return to an article solely to look for one of the references, whereas the humanists' footnotes are more convenient for those who want full details of the citation at the moment they read the text.

Because readers in different disciplines need to learn from each other, it will be increasingly necessary to accommodate such differences in usage and style. Toward this end, we feel it is important to publish citations in the manner used by each author.

Part 1
Introduction

How does contemporary research on neurotransmitters relate to our understanding of human behavior? What legal as well as scientific problems are posed by this new area of research? In addressing these questions, this volume will use findings concerning serotonin (5-hydroxytriptamine) to illustrate general problems that cut across disciplines in the natural sciences, social sciences, and law.

1

Why Study Serotonin, Social Behavior, and the Law?

Roger D. Masters

Abstract: What are neurotransmitters—and why should lawyers, social scientists, and informed citizens know about them? How does contemporary research in neuroscience relate to the problem of "nature-nurture"? Is the concept of legal or moral responsibility contradicted by the latest scientific research on the brain?

For the average citizen, as for most lawyers and scholars, it will seem paradoxical to connect law and biology. We usually think that legal rules are created by human decision, whereas biology, like fate, seems to be determined by external forces. Discoveries about the nature of the human animal over the last generation challenge these beliefs and pose an urgent problem. More specifically, as the work described in this book demonstrates, discoveries concerning the effects of neurotransmitters on human behavior obligate us to reexamine established ways of thought. Our freedom may depend on whether citizens and leaders have enough knowledge about advances in the biological sciences to control their use and their effects.

CONTEMPORARY BIOLOGY, HUMAN NATURE, AND LEGAL THEORY

According to conventional wisdom, evolution equals biology, biology equals genetics, and genetics equals determinism. Each of these equations is false. As a result, we need to abandon the widespread conception that a biological approach to human behavior somehow implies determinism and a loss of freedom.

3

•First, evolution does not equal biology. Evolution is the process of irreversible change and can be found in the history of languages and human cultures, as well as in other species. Biological systems evolve, but evolution is not limited to the things studied by biologists. And in the evolution of animal life, processes of change are often influenced by the behavior of the evolving animals themselves. The model of scientific determinism based on 19th-century physics—illustrated by the effects of gravity on falling bodies—does not apply to living things; evolution is often a complex process with unpredictable outcomes.

•Second, biology does not equal genetics. The life sciences study the interactions between genetic inheritance and environment. Biologists have discovered much about the way genes work, but genetics is only a small part of a discipline that includes ecology, ethology, zoology, cognitive neuroscience, botany, paleontology, and other subfields. Social behavior is particularly complex: what one particular animal does often depends as much or more on the behavior of other animals as it does on the individual's genetics and development. Biology as a whole is not reducible to molecular genetics.

•Finally, genetics is not determinism. Genetic processes follow predictable patterns, but these regularities concern the interactions of genotype and environment through the life cycle. The probable effects of a gene depend on the other genes, on the organism's development, and on the circumstances (including the social and physical environments). For humans, this means that our behavior influences the expression of inherited potentialities. Discovering the genetics of a human trait can—and usually does—*increase* our ability to control its effects.

In short, the theory and substantive findings of contemporary biology contradict the widespread opinion that a scientific perspective necessarily implies a denial of human freedom. The life sciences are about the history of the probable characteristics of populations of organisms, not about the predictable trajectories of falling bodies; Newtonian physics is no longer the relevant model of science (Mayr, 1985). When applied to humans, evolutionary perspectives help us to understand the uniqueness of our species (Alexander, 1990). In so doing, they explain why we have been able to control our own social behaviors and modify our environments to a degree unparalleled by other animals. Even more important, biological research can teach us how public policies that contradict human nature often have hidden costs, if not disastrous effects (Gruter, 1991).

The study of law is, of course, very different from that of biology. Law is more an art than a rigorous science. Legal history, constitutional law, jurisprudence, and the specific doctrines of the law do not seem to be integrated or explained in terms of a single generally accepted theory or philosophy of law. Lawyers have nothing equivalent to Darwin's theory or the neo-Darwinian "synthesis" that has become the overarching conceptual framework for the biological sciences. Laws are commonly established by the legislature and applied to individual cases by judges (and sometimes by juries) or by governmental agencies. These institutions develop conventional rules and procedures, which seem to depend on common sense (and sometimes on passion, bias, and even simple error). Different observers often disagree violently about what the law *is*, as well as what it *should be*. Nothing seems further from a natural science than the study and practice of law.

Appearances can, however, be deceiving. First of all, since law has a history, it can be said to evolve—and thus legal evolution can be compared to the evolutionary patterns studied in biology (Rodgers, chapter 13). Second, the *concept* of evolution was itself long a basic factor in American jurisprudence (Elliott, 1985); indeed, the divorce between biology and law that we take for granted emerged only in the last two generations for reasons that reflect politics and ideology rather than uncontested "truth" (Degler, 1991). Finally, but most important of all, legal institutions and behaviors are themselves biological phenomena in the broad sense that humans are living beings.

We evolved from primate ancestors over the last 10 million years and still share social behaviors, physical resemblances, and even genetic similarities with nonhuman primates. We are, of course, not the same as gorillas, chimpanzees, or baboons. But since we humans share over 90% of our genetic heritage with chimps, it is hard to think of our species as unique in all respects. To see how we differ from other animals, we also need to know exactly what elements are derived from our common primate origins (Schubert & Masters, 1991).

It is clearly unwise to think of human beings as merely apes without extensive body hair. Our enlarged brains, language, conceptual thought, complex societies, and the political, social, and technological inventions they have spawned, do make us unique in many regards. But these evolved traits were added to a primate repertoire rather than designed on a blank slate. Our primate heritage interacts with our culturally or individually learned acquisitions. In the broadest sense, legal and political behavior has to be seen as a highly unusual and extremely powerful adaptation of an unusual species rather than a total emancipation from natural processes and limits (Masters, 1989; Gruter, 1991). To think of humans as if we were *outside* of nature—as is implied by the common reference to "man's conquest of nature"—is neither accurate nor prudent.

This book illustrates the implications of a return to an older view that considered legal and political institutions in a naturalistic context. In the tradition that can be traced to Plato and Aristotle, humans are naturally social animals with a moral obligation to attend to justice and fairness (Masters & Gruter, 1992). This does not mean (if indeed it ever did) that the ideal or proper rules of behavior can be logically deduced from a static definition of human nature. Contemporary biology does not support the absolute and rigid definitions of natural law that are sometimes invoked in political controversy.

Environments change, bringing out different features of human potential. Societies evolve, developing different economic, social, and political institutions. But humans can no more ignore the limits of nature and human nature than Daedalus could fly merely by putting feathers on his arms and leaping into the air. What is needed, then, is a better sense of the constraints on future policy that might derive from our primate inheritance as it is expressed under contemporary circumstances.

THE HUMAN BRAIN, NEUROCHEMISTRY, AND BEHAVIOR

Generalities are not very helpful. Armchair speculation about human nature often focuses on abstract concepts like free will and determinism. In contrast, contemporary neuroscientists approach the human brain as an organ whose structure and functioning evolved from the mammalian central nervous system. They are not only discovering how we perceive, feel, and think, they are showing us how this scientific knowledge could increase our ability to control outcomes. It is time to look at some of the main points of this new way of thinking about human thinking.

The brain is not the blank slate or tabula rasa of philosophers like Thomas Hobbes and John Locke. Nor is it a computer made of identical units comparable to the silicon chip. The human brain, like that of mammals more generally and primates more specifically, has a structure that has evolved over millennia. This structure constrains how we sense the world, process that information, and act on it. In the light of research over the last generation, much of what happens in the brain is no longer a mystery.

As Paul MacLean (1983) has suggested, the human central nervous system can be understood as a system with three major structural components. The midbrain and brain stem, derived from the nervous systems of reptiles, control basic autonomic responses and the elementary movements that satisfy needs of locomotion, feeding, copulation, and the like. The cortex—the mass of "gray matter" at the top of the mammalian brain which expanded substantially among primates—is specialized in complex information processing; in humans, the primate cortex further developed to permit verbal language, complex memory, and the activities needed to read this book. The limbic system consists of structures between the midbrain and cortex, like the amygdala and hippocampus; their elaboration by mammals is associated with the emergence of emotion and the development of more complex learning and social behavior.

MacLean called this a "triune" view of the brain: the structures with a reptilian origin ("r-complex") governing simple consummatory responses, the limbic system functioning as the seat of emotions, and the cortex as the locus of complex thought. The parallels with Sigmund Freud's id, superego, and ego are obvious: the drives seem to arise from very primitive elements of the central nervous system, whereas the elemental emotions of guilt, fear, anger, and love need to be distinguished from the rational assessment and control of things around us and our own behavior. Such a view should hardly be frightening: long before Freud, Plato's "three parts of the soul" provided a similar image of human nature (*Republic*, IV).

Such a general conceptualization of the brain is, however, not very useful. The most recent advances in neuroscience have been based on a more precise localization of the way our central nervous system works. The current understanding has been described as the concept of a "modular" brain. Very specific functions are highly localized, and these localized units, or information-processing "modules," are linked in a complex, innate structure (Gazzaniga, 1985, 1988). The cells that allow our vision to discriminate a straight line differ from those discriminating curves; the visual processing of color is specialized in different cell structures from those perceiving shape, location in three-dimensional space, or movement. Visual pathways differ from olfactory or auditory pathways. Muscle control, producing the motor coordinations of each part of the body, is likewise localized.

The brain has evolved neuronal structures that link these various components in predictable ways. Vision is made possible by pathways that send impulses from the retina to the visual cortex in the back of the brain and thence to a variety of specialized information-processing modules. Visual perception must then be integrated with other sensory information. Memory and associative learning entail matching these inputs with stored responses—a process in which linkages between the cortex and the limbic system play a critical role. There even seems to be a specialized site in the left hemisphere of the brain, called "the interpreter module" by Gazzaniga (1985, 1988), which has the function of making "sense" out of apparently contradictory sense impressions.

The brain must therefore be seen as a complex system of linked structures. Neurons are not isolated cells forming a large array of identical computerlike chips. Localized assemblages of neurons, or modules, have to communicate with each other and, in so doing, actually change their structure and their interrelations. The brain is no more an unchanging organ than is the arm or leg: exercise, disease, and use shape neuronal structures and the way they function just as they shape our muscles. But some things are as impossible for our brains as was flying for Daedalus.

To understand how the system works, however, it is not enough to know the structure of the parts. The brain is an active organ. It requires constant intercommunication among distant neurons in order to integrate the activity of diverse modules. Here, it is neurochemistry that has revolutionized our understanding of the brain. But how does the process work?

Neurons are merely a number of different types of cells in the brain. When a neuron "fires," an electrical impulse is transmitted the length of the cell. For this to happen, a chemical event needs to occur at the junction (synapse) connecting one neuron with another. These chemical events, similar to fitting a key in a lock, are the focus of the study of neurotransmitters. Discoveries concerning the complexities of neurochemistry and their effects on behavior are accumulating at such a rate that this book is certain to be out of date before it is published. For that very reason, it is essential that the informed public, the legal profession, and the scholarly community know how it all works.

NEUROTRANSMITTERS AND HUMAN BEHAVIOR

Neurotransmitters and neurochemicals are of three broad types. Some, like serotonin, are principally known for their ability to trigger synaptic transmission; when they are present in the synapse (the space linking the two cells), impulses move from one neuron to another. The "receptor" site on the synapse that is triggered by such a neurotransmitter "fits" the shape of the molecule. As a result, different neurons are likely to be triggered by different neurotransmitters. Among the principal neurotransmitters known are serotonin, dopamine, and norepinephrine.

A second set of substances regulates the primary neurotransmitters. Some of these chemicals ("precursors") facilitate the body's production of a major transmitter. Others "compete" with a specific neurotransmitter, that is, chemicals with similar shapes that can fit on the same receptors and hence prevent the neurotransmitter from opening synaptic transmission (much as the "wrong" key of a similar make might block the "correct" key from opening a lock). Yet others bind to the neurotransmitter before it can contact the neuron and cause it to fire (like a free-floating lock searching out the key and latching onto it before it can open the door). And after neurons have fired, there are chemicals that prevent neurotransmitter decay, making the same molecules available for reuse in the brain.

Finally, a third set of substances—called "secondary messengers"—may be needed for the primary neurotransmitters to be effective. Here, the complexity arises at the level of the neuron being activated. Depending on the location in the brain, cells sensitive to a neurotransmitter like serotonin may need to be in contact with a second neurochemical along with serotonin before they fire. Hence in many cases there is the equivalent of a dual locking system such that the neuron will not fire unless both are opened at the same time.

Despite this variability, specific pathways in the brain seem to be particularly regulated by one or another of the major neurotransmitter systems. That is, some neuronal pathways may be primarily activated by dopamine and its associated neurochemicals, whereas others are principally controlled by the serotonergic system, norepinephrine, glutamate, and so forth. In some cases, however, a structure or pathway may be independently triggered by different neurochemical systems, as in sites that are sensitive to both dopamine and serotonin.

All of these complexities make it difficult to establish "constant" effects of any one neurochemical. But as the interrelations become known, the causal efficacy of brain chemistry appears to be increasingly important as a factor in everyday behavior, mental illness, and deviance. Many of the chemicals involved are modified by what we eat and drink or by what we do. A beer or a martini, no less than a tablet of aspirin or a shot of heroin, influences us by the chemical effects in the brain. As neuroscientists discover how the system works, it behooves the informed members of our society to become aware of the scientific findings and their implications.

BEYOND NATURE VERSUS NURTURE: CHANGING OUR CONCEPTS

Many readers will have assumed that the focus on neurotransmitters implies explanations based on "nature" rather than "nurture." Actually, the scientific perspective presented in this volume demonstrates that it is obsolete to think in these terms. In the light of contemporary biology, the so-called nature-versus-nurture controversy now appears to be meaningless: as Richard Alexander put it, to argue that something is caused by one *or* the other makes as much sense as asking whether the area of the rectangle is caused by the length or by the width.

Consider the range of factors now known to influence human behavior either by modifying the structure of the brain or by modifying the activity of the neurotransmitter systems that regulate its function. Studies of heritability and biochemistry have suggested that there are genes that increase the risk of specific mental illnesses (such as schizophrenia) or behavioral predispositions (such as panic disorder or extreme shyness). But even here, the effect of such genes need not be automatic since sometimes they depend on specific experiences or traumas to trigger the genetic effect.

The nature-nurture dichotomy is even more directly contradicted by prenatal effects on physiology and behavior. Is fetal alcohol syndrome a question of nature because the affected child is born with cerebral malformation and cognitive deficits? Is it a matter of nurture insofar as the mother's consumption of alcohol is responsible? How does one sort out the contribution of cultural deprivation or socioeconomic disadvantage, which leads some women to use alcohol or drugs (particularly if the social environment leads to abandonment by a fetus's father), and the biochemical properties of alcoholism (which may be a tendency that is to some degree genetic)? And, among Native American populations, the absence of the gene that breaks down ethanol among Caucasians makes the consumption of alcohol far more dangerous: since this "biological" factor interacts with cultural isolation to produce a particularly great incidence of fetal alcohol syndrome on Native American reservations, the attempt to reduce explanations to *either* nature *or* nurture merely produces confusion and misinformation.

Even the factors traditionally described as environmental influences on behavior are more diverse than has typically been admitted. Usually, the nature-versus-nurture dichotomy is used to explain the importance of culture or socioeconomic status. But with regard to many elements of human behavior, the main sources of variation are either within-family variation (the differences in experience between siblings) or inheritance. Studies of personality show, for example, that culture and economic or social class have little to do with the difference between shy and gregarious behavior or the incidence of various forms of mental illness. Rather, as mothers have long observed, it is the microenvironmental difference between brothers and sisters—and not the status of the entire family—that explains differences in these traits.

Culture does, of course, also influence behavior. In particular, we are often influenced by what we eat (or drink), as well as by what we do or what we think. But while cultural practices may have an effect on behavior, these effects are sometimes very much related to neurochemistry—as is illustrated by the spread of drug use in the United States. Humans only consume alcohol and other psychoactive compounds because of chemical effects on mood and behavior. And not everyone has the same reaction to any one substance, be it crack-cocaine or whiskey. If nothing else, the neurochemical revolution, by explaining the role of chemistry in our brains, requires that we abandon the overly simplistic concept of a dichotomy between nature and nurture. To show why this is true, consider the different ways that neurochemistry can affect us.

NEUROCHEMISTRY, PERSONALITY, AND BEHAVIOR

Neurotransmitters and hormones can have effects of at least four types: (1) long-term baselines that establish differential probabilities of behavioral responses; (2) "resetting" of a baseline in response to environmental or social change; (3) daily or monthly cycles; (4) very short-term changes in response to specific events. In addition, the effects of any one neurotransmitter often seem to depend on interactions with other compounds that regulate the relevant neurotransmitter system *and* with the other neurotransmitter systems as a whole.

This complexity can be illustrated with regard to serotonin. Functionally, serotonin seems to be associated with a number of different effects. Although originally known for its role in the digestive process, serotonin is a neurotransmitter that coordinates different muscular responses in simpler organisms. In many sites in the brain (especially in the sensory pathways), it is a major inhibitory neurotransmitter: hence individuals with lower levels of serotonin often exhibit deficits in impulse control. Across species ranging from lobsters to vervet monkeys, dominant individuals exhibit higher levels of serotonin. Behaviorally, higher serotonergic activity is often associated with more controlled or smoother motor coordinations; lower levels may coincide with uncontrolled outbursts of aggression, especially directed toward "inappropriate" targets (e.g., an adult male vervet attacking an infant without provocation).

Since the effects of any one neurotransmitter seem to be, to an extent, localized and functionally specialized, variations in serotonin activity seem to be associated with specific kinds of behavior that differ from the responses associated with levels of dopamine or norepinephrine. Individuals seem to differ in baseline activity of the pathways in the brain that are composed of neurons for which serotonin is a principal neurotransmitter (so-called serotonergic pathways). But behavioral responses, or personality traits, do not depend on any one transmitter alone; as a result, individuals with exceptionally low baseline levels of serotonin may—depending on the role of other neurotransmitter and neurochemical systems—be depressive, suicidal, homicidal or impulsively aggressive, subject to seasonal affective disorder, or normal.

While some effects of serotonin concern its baseline effects (which may well be to some degree heritable), others involve situation-specific "resetting" of serotonergic activity. Here the work of McGuire, Raleigh, and their collaborators (chapter 9) is essential: the behavioral correlates of high status seem to produce, over a 14-day period, an elevation of whole-blood serotonin (WBS), whereas loss of status or isolation entails a reduction of serotonin levels over a similar period. Since these changes are triggered by behavioral interactions but can be simulated by neurochemical treatments of the animal, the effect seems to be of a different order than the baseline effects mentioned above. Because, as Sapolsky (1990) has shown, winning dominance encounters also has short-term effects on testosterone levels, primate social behavior is regulated by neurochemical reactions of different types.

These issues may seem technical to the layman, but they can be readily translated into more familiar terms. The baseline levels of neurotransmitter activity seem to correspond to what we call "personality" (i.e., relatively stable dispositions to respond to the social and natural environment in predictably different ways). The short-term "reset" of baseline in response to changes in social status is a different effect, comparable to what we might call a "behavior change" (analogous to what are conventionally described as "attitude changes"). This distinction is important because some of the serotonin effects reported in this book seem to be associated with personality rather than with changes in the status or dominance of the individual.

The scientific definition of personality or "temperament" has been the subject of considerable controversy. For many years, since the pioneering work of Eysenck, some personality theorists have proposed classifications based on three distinct dimensions. Robert Cloninger (1986, 1987) has recently proposed a theory of personality based on three factors that are in turn presumed to derive from differences in the activity of three major neurotransmitter systems: those controlled by dopamine, by serotonin, and by norepinephrine (see also MacDonald, 1988).

This theory is still highly tentative and will certainly be revised in the light of future discoveries. Although other neuroscientists have somewhat different but comparable models of the way neurotransmitters influence human behavior (cf. Eaves, Eysenck, & Martin, 1989), it will be useful to describe the approach of Cloninger and MacDonald as an illustration of how neurotransmitter systems might be related to personality. Even if only approximate and revised by future research, the broad outlines of this type of analysis seem likely to characterize the way chemistry can contribute to the enormous diversity of human responses to any one situation. As Madsen's work (chapter 10) makes clear, these effects are particularly important in human behavior.

Each of the three neurotransmitter systems identified by Cloninger and MacDonald is associated with a different dimension of personality. Cloninger suggests that serotonin is linked with "harm avoidance" (the propensity to avoid or to take risks). Norepinephrine is, in this view, associated with "reward dependence" (the tendency to seek and be sensitive to social reward as the mediator of behavior). According to Cloninger, dopaminergic levels are related to "novelty seeking" (the tendency to search for novel stimulation), though MacDonald suggests that this dimension might better be viewed in terms of emotionality. Since each system influences the action of the others, any one individual needs to be analyzed in terms of his or her activity levels on all three personality dimensions (or all three neurotransmitter systems).

According to Cloninger's theory, the serotonergic system is associated with harm avoidance and risk taking: low levels of serotonin are associated with the tendency to avoid risk and danger, whereas high levels are linked with a greater propensity to take risks. It is not necessary to accept this conceptualization of personality in all respects to see its relevance to commonly observed differences in the way people behave.

It is interesting to note that something like these personality dimensions has been found in other mammals. Among vervet monkeys, for example, the same individuals were observed in socially varied contexts. Analysis of descriptions of their behavior revealed individually varied patterns of response that were to some degree constant across situations; these continuities reflect three factors which resemble the personality dimensions distinguished by Cloninger (McGuire, personal communication). It is thus worth considering how neurotransmitters like serotonin might influence behavioral dispositions and individual character traits like those described as personality by psychologists.

Increasingly, neurotransmitters are being understood as "tuning" devices that modify the signal-to-noise ratio, or "gain," of specific tracts in the cortex (Servan-Schriber, Printz, & Cohen, 1990). In practical terms, this means that different levels of a neurotransmitter may have quite specific effects in modifying sensitivity or amplitude of response to specific kinds of social cues of threat, reassurance, sexual attractiveness, and the like. This might be called a "rose-colored-glasses" theory of personality. For example, given the inhibitory effects of serotonin, individuals with low levels of this neurotransmitter might see threat in ambiguous cues that are ignored by those with higher levels of serotonin. As a result, the serotonergic baseline differences would translate into a different probability of seeing a particular social situation or cue as threatening or not.

To see how this might work, consider the tendency of some individuals to be aggressive, manipulative, and socially devious. Psychological measures of this character trait, like the so-called Machiavellianism scale (Christie & Geis, 1970) can be seen as a form of personality test, measuring the individual's baseline tendencies or propensities in social behavior. This scale seems related to both what Cloninger calls reward dependence and harm avoidance: the more Machiavellian an individual, the lower that person would score in measures of the dependence on social reinforcement or reward *and* the lower that person would score in the tendency to avoid harm or risk taking. Whereas aggressiveness is sometimes described as a kind of dominance, the personality traits involved in social manipulativeness would thus concern behavioral dispositions associated with baseline levels of neurotransmitter activity.

In this case, note that at least two neurotransmitter systems would be involved (those associated with serotonin and norepinephrine). Each neurotransmitter system, moreover, is dependent on a number of different neurochemicals, and each is sensitive to resetting by experience. Hence measure of a single neurotransmitter would tell us little or nothing about an individual. And even if we can say something about personality dispositions, the individual's experiences and social situations can change the baseline and produce lasting modifications in response. As a result, even if personality could be traced in part to neurochemical processes, the results would only concern differing degrees of sensitivity or potentiality, explaining what is generally known as "character" without denying the role of individual learning and responsibility.

Some studies have shown that individual differences in response to a specific social situation can work in practice. Jerome Kagan and his colleagues, for example, have shown experimentally that some children are exceptionally shy, while others are unusually gregarious. At the extremes of the dimension of sociability, children do not change much depending on experience between the ages of 2 and 7; individuals nearer the average, on the other hand, are more influenced by their personal experiences (Kagan, Reznick, & Snidman, 1988). But however the child's baseline of social response is established, the individual has a propensity to respond to a particular social situation that is reflected in physiological responses of heart rate, stress hormones, and the like (cf. Montagner, 1978).

While differences from one individual to another often depend on changes in social status, at any moment each of us has a particular set of social predispositions. These traits, which we call personality, influence—and in turn are influenced by—our thought processes. In studies of viewer responses to televised images of leaders (Masters, 1989; Sullivan & Masters, 1988), subjects have been found to differ in their sensitivity to cooperative and competitive cues. That is, controlling for all cognitive and emotional factors, there are significant differences in individual sensitivity to aggressive or reassuring cues, as would be predicted by this theory. A first exploratory analysis shows, moreover, that subjects' scores on Cloninger's personality dimensions seem to predict sensitivity to these different classes of social stimuli. In short, even in such day-to-day behavior as watching leaders on television or responding to the social cues of others, neurotransmitter levels and activity may be a powerful reason for differences in perception, sensitivity, and behavior.

WHAT DOES IT MATTER? SEROTONIN, DEVIANCE, AND MENTAL ILLNESS

The effects of neurotransmitters like serotonin on personality and day-to-day behavior are still the subject of scientific debate. As always in the history of science, new findings, hypotheses, and theories are controversial. Nonscientists would, however, be ill advised to conclude that disagreements among experts are a sign that nothing of practical importance is known as yet about the role of neurotransmitters.

Discoveries in at least two areas are particularly important for decisions concerning law and public policy. The first is deviant or criminal behavior; the second, forms of mental illness (including suicide). In both areas, even though studies in progress do not yet give us definitive answers (or rather, precisely for this reason), neuroscientific research indicates why it is now necessary to rethink old approaches.

Because serotonin is often involved in the inhibition of neuronal response, individuals with low baseline activity of some serotonergic pathways seem to be more likely to experience deficits of normally inhibited reactions. Depending on the activity of other neurotransmitter systems, this can take the form of impulsive violence (such as homicides in which the name of the victim is not known), suicide (violence directed at the self), or depression (inability to inhibit thinking about negative or threatening experiences and outcomes).

At the extreme, as will be seen in chapter 6 by Linnoila and his colleagues, it may be possible to predict future criminal behavior on the basis of biochemical assays. Even if one contests the adequacy of their data, the mere existence of research on these matters necessarily affects our legal system. Can a defense lawyer invoke low levels of serotonin as a mitigating circumstance in a criminal case? How should information about neurotransmitters be integrated into our understanding of crime? And, perhaps even more important, how might it be related to punishment or to treatment of the criminal? Even if not admitted as evidence in the criminal case itself, isn't such information relevant to decisions concerning probation or the medical treatment of prisoners?

Civil law is perhaps even more important than criminal law. If neurotransmitters are implicated in suicide and depression, who should "own" the information about an individual's biochemistry? Are physicians obligated to communicate results of laboratory tests to patients' families, prospective spouses, or employers? How will insurance companies integrate these findings into the terms of their policies? At what point does our society need to start teaching its citizens what is scientifically known about our brains and neurochemistry?

These and many other questions like them are likely to become matters of public debate in coming years. How, then, can we best prepare to assess new discoveries in a way that is consistent with traditional values of privacy, individual autonomy, and social decency? What are the implications of decisions to fund highly technical scientific research? And, how might legal decisions have quite unanticipated consequences in changing environments? Knowledge of the complexities of both scientific and legal dimensions can only assist our society in coming to grips with the often unsettling discoveries that are the subjects of parts 2 and 3 of this book.

For example, it is suggested in the concluding chapter that research on serotonin should not be perceived primarily in terms of criminal law. Instead, there may be important advantages in considering abnormal levels of serotonergic functioning as providing an entitlement to special education in the early years of childhood. Instead of leaping to the obvious implications about criminality, perhaps an individual's level of neurotransmitter activity might best be viewed as private medical information that is principally relevant to claims for compensatory education. Here, the model might well be dyslexia, another area in which neurological discoveries have influenced our social and legal system by establishing the concept of entitlement to educational services to help individuals compensate for the deficit (see chapter 16).

In coming years, lawyers, scholars, legislators, and informed citizens will increasingly be called upon to deal with issues that require informed discussion. The way we conceptualize scientific information and its relevance to public policy is not a given. If the general public is not kept abreast of biological discoveries that at first seem highly technical, experts may have hitherto unknown opportunities for the misuse of power. Without a concerted effort to relate neuroscientific research, law, and the traditional approaches to human behavior, it is hard to see how our constitutional system could to survive a future epoch in which some might have the technological ability to control the thoughts and behaviors of others through chemical means.

CONCLUSION

Although more work needs to be done, the current volume is a first step in understanding the complex interrelations between neurochemistry, personality, and social behavior. It suggests, however, that the complexities found in the empirical studies are derived from differences in baseline behavioral potentiality, that is, that basic differences in neurochemical activity are associated with what we call personality rather than with a specific behavior itself. And since such personality differences seem to be partially heritable but subject to readjustment due to individual experience (e.g., Kagan, Reznick, & Snidman, 1988; Eaves, Eysenck, & Martin, 1989; Plomin, 1990), the complex interactions between nature and nurture may be illuminated by further research and dialogue concerning the findings reported here.

The remainder of part 1 sets out the scope of the problems at issue. In chapter 2, McGuire describes the relationship between law and biology in a general way; in chapter 3, I then summarize the empirical studies that make up the rest of the book. Part 2 outlines a number of areas in which serotonin and serotonergic activity influence behavior: after outlining the basics of the neurochemistry (chapter 4, Yuwiler, Brammer, and Yuwiler), examples are given concerning suicide (chapter 5, Stein and

Stanley), impulsive behavior (chapter 6, Linnoila and coworkers), and seasonal affective disorder (chapter 7, Wurtman and Wurtman). Part 3 turns to factors that influence serotonergic function, including social experience (chapter 8, Ginsburg), social status (chapter 9, Raleigh and McGuire), and individual personality (chapter 10, Madsen). Part 4 turns to the way these findings challenge legal concepts, focusing on the proposal that crime be treated as a medical condition (chapter 11, Jeffery) and the effect of neurochemistry on our concept of legal responsibility (chapter 12, Shapiro). The book concludes, in part 5, with a survey of the implications for law and public policy, beginning with the way evolution needs to be linked to legal change (chapter 13, Rodgers) and turning to the therapeutic value of criminal law (chapter 14, Wexler), the impact of expenditures on scientific research (chapter 15, Goldberg), and overall conclusions with regard to public policy (chapter 16, Masters).

REFERENCES

Alexander, Richard D. 1990. How Did Humans Evolve? Reflections on the Uniquely Unique Species. *University of Michigan Museum of Zoology*, Special Publication No. 1, Ann Arbor.

Christie, R., and Geis, F. 1970. *Studies in Machiavellianism*. New York: Academic Press.

Cloninger, C. Robert. 1986. A Unified Biosocial Theory of Personality and Its Role in the Development of Anxiety States. *Psychiatric Developments* 3:167–226.

———. 1987. A Systematic Method of Clinical Description and Classification of Personality Variants. *Archives of General Psychiatry* 44:573–88.

Degler, Carl. 1991. *In Search of Human Nature*. New York: Oxford University Press.

Eaves, L. J., Eysenck, H. J., and Martin, N. G. 1989. *Genes, Culture, and Personality*. New York: Academic Press.

Elliott, E. Donald. 1985. The Evolutionary Tradition in Jurisprudence. *Columbia Law Review* 85:38–94.

Gazzaniga, Michael. 1985. *The Social Brain*. New York: Basic Books.

———. 1988. *Mind Matters*. New York: Norton.

Gruter, Margaret. 1991. *Law and the Mind*. Newbury Park, CA: Sage Publications.

Kagan, Jerome, Reznick, J. Steven, and Snidman, Nancy. 1988. Biological Bases of Childhood Shyness. *Science* 240:167–71.

MacDonald, Kevin. 1988. *Personality and Social Development*. New York: Plenum.

MacLean, Paul. 1983. A Triangular Brief on the Evolution of Brain and Law. In M. Gruter and P. Bohannan, eds., *Law, Biology, and Culture*, pp. 74–90. Santa Barbara, CA: Ross Erikson.

Masters, Roger D. 1989. *The Nature of Politics*. New Haven: Yale University Press.

Masters, Roger D., and Gruter, Margaret, eds. 1992. *The Sense of Justice*. Newbury Park, CA: Sage Publications.

Mayr, Ernst. 1985. How Biology Differs from the Physical Sciences. In Depew and Weber, eds. *Evolution at a Crossroads*, pp. 43–63. Cambridge, MA: MIT Press.

Montagner, Hubert. 1978. *L'enfant et la communication*. Paris: Stock.

Plomin, Robert. 1990. The Role of Inheritance in Behavior. *Science* 248:183–88.

Sapolsky, Robert. 1990. Stress in the Wild. *Scientific American* 262:116–23.

Schubert, Glendon, and Masters, Roger D., eds. 1991. *Primate Politics*. Carbondale: Southern Illinois University Press.

Servan-Schriber, David, Printz, Harry, and Cohen, Jonathan D. 1990. A Network Model of Catecholamine Effects: Gain, Signal-to-Noise Ratio, and Behavior. *Science* 249:892–95.

Sullivan, Denis, and Masters, Roger D. 1988. "Happy Warriors": Leaders' Facial Displays, Viewers' Emotions, and Political Support. *American Journal of Political Science* 32:345–68.

2

Biology and the Law

Michael T. McGuire

Abstract: How do biological theories and research findings relate to the law? Recent changes in both biology and law put the "neurotransmitter revolution" into a broader perspective, emphasizing the need for informed dialogue between specialists of diverse disciplines.

Over the last four decades, biology has significantly changed our understanding of human behavior, and continuing changes can be anticipated through the 1990s. During this period, law has assimilated only a part of what biology has discovered. To say this is not to criticize. Rather, it is to identify a *gap* that separates the two fields, a gap that requires analysis, explanation, and, perhaps, rectification.

BACKGROUND

At any given time in the history of a society, the existing body of laws and the degree to which laws influence behavior can be thought of as the product of interactions between groups with different needs, strategies, values, and understanding. Such differences are apparent in the importance individuals place on

 •*Tradition and Culture*, in the sense that prior cultural values, family values and living styles, experience, and ideals are incorporated into and addressed by law;
 •*Reason*, in the sense that interpretations and valuations of events and behavior, as well as the social reality in which events occur, influence and are addressed by law;

17

•*Needs and Expectations,* in the sense that individuals' feeling states, economic and personal needs, and expectations of their own and others' social behavior influence and are addressed by law;

•*Human Nature,* in the sense that general species-characteristic patterns of behavior and physiological responses to events influence and are addressed by law;

•*Environment,* in the sense that the access to, and the ownership, use, and consequences of use influence and are addressed by law.

The preceding list is far from exhaustive, and alternative categories could be developed (e.g., individual and social freedoms). The *needs and expectations* category might be subsumed within the category designated *human nature,* but there are obvious influences of *tradition and culture* on needs, and *needs and expectations* in part reflect environmental options. There are also law-influencing, rule-making groups (e.g., legislatures, lobbyists), and rule-interpreting groups (e.g., courts, environmental groups), which influence and are addressed by law. These groups have different values, internal dynamics, and practices that affect the pace at which new knowledge is incorporated into law, as well as the form it takes (see, for example, Rawls, 1971; Pepper, 1970).

Contemporary legal and social theories often simplify the understanding of human events, in extreme cases reducing the categories outlined above to a dichotomy between "nature" and "nurture." In jurisprudence, this has taken the form of legal positivism: laws are seen as the result of human will, convention, or stipulation; whereas human nature and the physical environment reflect a totally different order of causes and events (Murphy, 1990). In the social sciences as a whole, a similar division between human agency and biology has long dominated theory and research. This distinction between human law, or culture, and human nature, or biology, can no longer be accepted uncritically. Biological research has transformed our understanding of the causes of human social behavior, making it impossible to defend a simplistic nature-nurture dichotomy on scientific grounds.

Historically, persons or groups advocating the importance of the categories listed above have exerted their influence on legal philosophers, legislators, and practicing judges and lawyers (Murphy, 1990; Gruter, 1990). As a rule, interactions among *human nature, tradition and culture,* and law are continuous over time, although the frequency and intensity of interactions vary. There exist recently enacted laws that grant special rights to pregnant women and that require protective clothing in hazardous jobs. However, law has been slow to assimilate information about the hazardous effects of certain drugs on automobile driving and the adverse impact of stressful working conditions on health. New laws are also visible with respect to the *environment.* Such laws constrain land use in areas in which there are endangered species and prohibit acts that have long-range negative environmental impacts, such as certain types of waste-disposal practices. *Needs and expectations* serve both to initiate and to impede lawmaking efforts, as well as to influence the degree to which individuals comply with existing laws. The changing fortunes of antitrust laws and the degree to which they are enforced are examples. The types of interactions described above are selectively interpreted and reacted to by members of rule-making and rule-interpreting groups. Out of this political, frequently conflictual, and often seemingly directionless set of interactions and events come our laws, some of which are sensible and work and some of which are not so sensible and do not work.

Biology is an increasingly important factor with respect to understanding human behavior and the sensibility and workability of laws (Masters, 1989; Gruter, 1990). Examples since the 1950s include the identification of (1) heritable predispositions for species-typical social behaviors, such as the strong motivation to acquire and control resources, to defend territories, and to engage in nepotistic and sexually possessive behavior (see Barash, 1982); (2) genetic contributions to vulnerability for traits of mental illness (once explained as entirely due to social factors) (Kaplan and Sadock, 1988) and criminal behavior (Brown et al., 1990); (3) consequences of exposure to chemicals, such as asbestos fibers, drugs, and altered foods, including prenatal developmental disturbances or traumas that produce irreversible physiological and neurological damage (e.g., fetal alcohol syndrome); (4) physiological consequences of social stress, particularly on cardiovascular and psychological function (Goldberger and Breznitz, 1982) and lasting deleterious physical and behavioral consequences of child neglect and child abuse and other disruptions of the normal developmental process; (5) interactions between environmental and heritable factors in producing various forms of depression, premenstrual syndrome, and criminal behavior; (6) the long-term impact of certain types of environmental destruction and reduction of species diversity (Wilson, 1984); and (7) control over the reproductive process, including contraception, pregnancy interruption, and prenatal screening and diagnosis.

Findings like those noted above inform law about the causes of behavior, the inevitability of certain behaviors or physiological responses given particular conditions, the degree to which compliance with specific laws is likely, and the probable impact of different physical and social environments on nonhuman life-forms. In many instances, such information had not previously been known or, if it had been, its implications had not been appreciated.

The manner in which law is informed, as well as the way in which biological information is utilized by law, is far from consistent. Biological information about the consequences of specific working conditions is often precise and rapidly incorporated into law. Knowledge about the long-range effects of reducing species diversity is less precise, and, generally, those actions that are designed to protect species are more slowly incorporated into law (and are more frequently opposed because of economic interests). Yet information about the highly probable behavioral characteristics of human nature has the most unpredictable fate of all. Consider, for example, deception for the purpose of acquiring resources (e.g., stock-market tips, inadequate disclosure of known economic facts, and disregard of environmental laws). Not only is such behavior frequent (see Mitchell and Thompson, 1986, for a review), but findings from biology strongly suggest that deceptive behavior, used for the purpose of personal gain, is strongly predisposed and unlikely to change. One may ask: To what degree do our current laws take these findings into account?

Another and important side to the biology-law relationship focuses on the effects of technological developments. For example, biologists and engineers have developed drugs that alter behavior, techniques of sperm storage and transfer, and technical means of prolonging life. Each of these developments impacts human behavior in one way or another. How the law might optimally deal with these developments results in an increasingly complex set of questions. For instance, under what conditions should behavior-altering drugs be administered against the will of individuals? How should limited and costly medical resources that prolong life be allocated? What costs should industry assume to reduce harmful industrial processes? And, in a lighter vein, should the father of an anonymous sperm donor have visitation rights to his genetic grandchild?

There are neither easy nor quick answers to these questions. Moreover, however compelling biological findings may be to biologists, biological information enters a complex intellectual and value-influenced universe in which the knowledge, interests, and values of biologists, legal scholars, lawyers, politicians, legislators, business executives, and religious and ideological groups are often in conflict.

An instructive example of the complexity involved in dealing with available biological information is provided by studies that show that the activities of certain hormones and neurotransmitters change as a consequence of social interactions. The principle involved is well known in the area of stress research, where studies have focused on relationships between stress, changes in "stress hormones" (e.g., cortisol), and (generally) negative cardiovascular effects. Yet similar events occur with respect to a number of nonstress hormones and neurotransmitters. Negative feedback, for example, significantly alters serotonin activity. Absence of feedback alters prolactin and growth-hormone activity, as well as norepinephrine activity. Positive feedback alters testosterone activity, and so on (see McGuire and Troisi, 1987, for a review). Increases in stress alter one's health state and life expectancy. In the short term, changes in the activity of selected neurotransmitters alter, for example, how a person processes information (i.e., interprets the world), the probability that one will act impulsively or aggressively towards others (see McGuire, 1986, for a review), and the chances that one will withdraw socially. If such changes persist, severe mental disorders (psychoses) may develop, or suicide may occur.

There are, of course, more obvious examples of complex situations resulting from new biological information (e.g., whether persons with certain types of communicable diseases need to reveal this information in applying for jobs). Generally, such information quickly finds its way into the law. Such speed can be contrasted with examples from the preceding paragraph, which were selected because they are often overlooked by the law. How the law might deal with available information about the consequences of behavioral interactions is addressed in subsequent chapters. For this chapter, the critical point is: it is likely that findings from biology and medical research will become increasingly compelling and will come to influence legislative and management approaches of persons who engage in certain behaviors and whose behaviors are a consequence of factors outside their control. Given this possibility, two questions arise: what kind of information is likely to be incorporated into the law; and within what time frame might incorporation occur?

Before trying to answer these questions, one additional point is worth noting, that of law's influence on biology. Just as biology informs law, law also informs biology, such as in environmental law, where laws are enacted to restrict free use of the environment prior to a full biological understanding of the consequences of such laws: we do not know how many nonhuman species might be endangered if a particular ecological niche is destroyed in the process of developing a housing project. Or consider the relatively new technological area of genetically altered biological materials. It has been necessary to establish rules and, in some instances, laws for the manufacture and distribution of such materials far in advance of our understanding of the full environmental and health implications of many of these alterations. Such laws frequently stimulate new efforts by biologists to try to answer previously unstudied questions.

Returning to the questions that deal with what kind of information is likely to be incorporated into the law and the time frame of incorporation, perhaps the first point to note is that law generally adopts a conservative posture with respect to change and the assimilation of new information. There are multiple reasons for this conservatism, many of which make good sense. As this point applies to findings from biology, perhaps the most defensible reason is that it requires time to resolve conflicting interpretations and valuations of biological information. Biological findings, however exact and consensually agreed upon, can seldom be incorporated into the law without prior translation from a technical language to a language usable in law. Translation inevitably results in ambiguities. This situation is not necessarily the fault of biologists. Relative to many other species, *Homo sapiens* is a highly variable species; and experiences, expectations, biological states, and needs differ not only across persons but the intensity and frequency with which they influence behavior also varies. The specificity of findings must also be considered. Reduced central nervous system serotonin activity, for example, is observed in a number of conditions unrelated to impulsive behavior, such as depression, and as a statistical variant in the normal population. Similar points apply to values and to the importance accorded tradition— people value past rules, mores, and living styles differently. Further, law has purposes other than immediately attending to each new potentially relevant piece of biological information. The attempt to uphold and create moral and ethical values, that is, filtering consensually validated reality through existing norms to address social expectations, is one such purpose. Molding behavior, or bringing behavior into conformity with ideological norms, is another. Testing the usefulness of existing laws is yet another. In a social environment best characterized by multiple and often conflicting values, forces, and interests, change is understandably slow and this is, on balance, wise.

There are consequences, nevertheless, and one is that, on average, only a limited number of relevant biological findings will be incorporated into law over any given period of time. Exceptions exist in situations where legislative bodies dramatically respond to information about such things as drugs that damage fetuses; irrational behavior, such as random freeway shootings; or the presence of newly discovered environmental contaminants, such as radioactive dump sites beneath grammar-school playgrounds. Responses are especially dramatic when there is both compelling biological evidence dealing with the harmfulness of particular behaviors (random shootings) or substances (radioactive waste materials), as well as a strong desire among the public for change. Yet dramatic changes in law remain the exception.

Far less dramatic timetables prevail for information that is not definitive or for behavior that appears to be difficult to influence by law. Examples include situations (1) where biological evidence is highly suggestive but not irrefutable; (2) where biologically predictable deceptive behavior, presumably under the control of individuals, may be socially disadvantageous (illegal business deals); (3) where only a small percentage of the population is affected, as in environmental safety standards in certain low-profile, small industries (e.g., the pulmonary consequences of brick dust among persons working in brickyards), perhaps also to the "rights" of the homeless; and (4) where a large percentage of the population is implicated yet is not sufficiently well organized to alter the law, a point that applies to continuing racial inequality and to unequal pay for women.

The recognition and acceptance of relevant biological knowledge would not necessarily bring an end to many of the conflicts that currently characterize public and legal arenas. Moreover, there are many events that biology can explain only in principle but not in detail, which is an essential ingredient of law. Specific cultural values are examples. An obvious conclusion emerging from studies by comparative legal scholars is that different societies deal with essentially the same biological information in different ways. Nevertheless, it is hard to make a compelling argument that biological information should not be incorporated into law or that it should be incorporated in a haphazard way. Systematic incorporation would certainly result in a reformulation of many of the legal, philosophical, and private-interest positions currently influencing law, as well as set the groundwork for more sensitive and biologically responsive laws.

In bringing this chapter to a close, I identify two kinds of activities that I believe would lead to more productive interactions between biology and law within the context of current social reality. First, from the perspective of biology, there is a need to evaluate systematically and routinely new information and to formulate it in ways such that its implications are understandable to nonbiologists. I would hasten to say that, in my view, this recommendation would not be easily accomplished. Biology is organized around individual scientists and their theories and findings, and they cooperate only in certain situations. At least this is largely true in the United States. A new social-philosophical perspective would be essential if biologists were to engage routinely in such activities. Nevertheless, the systematic and routine evaluation of biological knowledge needs to be accomplished.

Second, from the perspective of law, there is a need to evaluate its internal processes by which the assimilation and evaluation of new biological information takes place. Apart from whatever constraints (e.g., lobbying, education) might exist in accomplishing this goal, the task itself is nontrivial. It will therefore be necessary to train a new group of lawyers and social scientists so that they become aware of the developments in the life sciences, as well as of the complexities of our legal and political system.

Most important, there is a need for a continuing and active dialogue between biologists, legislators, and lawyers. Such a dialogue may help prevent the existing gap from widening.

ACKNOWLEDGMENTS

Margaret Gruter, Terrance McGuire, and Roger Masters all provided comments for this chapter, and their help is appreciated.

REFERENCES

Barash, D.P. *Sociobiology and Behavior*. New York. Elsevier. 1982.

Brown, G.I., Linnoila, M., and Goodwin, F.K. Clinical assessment of human aggression and impulsivity in relationship to biochemical measures. In H.M. van Praag, R. Plutchik, and A. Apter (eds.), *Violence and Suicide*. New York. Brunner/Mazel. 1990. pp. 184–217.

Goldberger, L., and Breznitz, S. *Handbook of Stress*. New York. The Free Press. 1982.

Gruter, M. *Law and the Mind*. Thousand Oaks (CA). Sage. 1990.

Kaplan, H.I., and Sadock, B.J. *Synopsis of Psychiatry*. Baltimore. Williams and Wilkins. 1988.

McGuire, M.T. Biochemical screening to predict behavior. *Univ. Southern Calif. Law Rev.* 1991.

McGuire, M.T. (ed). *Health-Behavior-Disease*. Kalamazoo (MI). The Upjohn Company. 1986.

McGuire, M.T., and Troisi, A. Physiological regulation-deregulation and psychiatric disorders. *Ethology and Sociobiology*. 8:95–125. 1987.

Masters, R. *The Nature of Politics*. New Haven. Yale Univ. Press. 1989.

Mitchell, R.W., and Thompson, N.S. (eds.). *Deception*. Albany. State Univ. New York Press. 1986.

Murphy, J.B. Nature, custom, and stipulation in law and jurisprudence. *Rev. Metaphysics*. 43:751–790. 1990.

Pepper, S. *The Source of Values*. Berkeley. Univ. Calif. Press. 1970.

Rawls, J. *A Theory of Justice*. Cambridge. Harvard Univ. Press. 1971.

van Praag, H.M., Plutchik, R., and Apter, A. (eds.). *Violence and Suicide*. New York. Brunner/Mazel. 1990.

Wilson, E.O. *Biophilia*. Cambridge. Harvard Univ. Press. 1984.

3

Serotonin and the Law

Roger D. Masters

Abstract: This chapter provides a synopsis of the book, outlining the findings in neurochemistry and their implications for law and public policy.

Knowledge is a dangerous thing. Research on the life sciences has advanced our knowledge of the structure and function of the human brain beyond anything generally expected 100 years ago. We now have an understanding, at least in outline, of the contributions of genetics, physiology, and individual experience in establishing constraints on behavior. For a number of psychological conditions, this knowledge makes it possible to change the mental state and behavior of individuals through either environmental manipulation or psychoactive drugs. And, in many ways, this information is transforming the meaning of human choice and responsibility.

It is, therefore, timely to explore the legal implications of recent research in behavioral neurochemistry, with specific attention to the effects of serotonin on the social behavior of humans and nonhuman primates. This volume cannot, of course, claim to be definitive: on the contrary, research is advancing so rapidly that any such effort would be doomed to futility. Rather, we hope to illustrate the kind of interdisciplinary dialogue that is necessary if the "neurotransmitter revolution" is not to have disastrous effects for our legal system and our society as a whole. To this end, it is necessary to combine perspectives from many disciplines. Lawyers and criminologists need to engage in dialogue with life scientists who study neurochemistry and its effects. Political scientists or philosophers may also contribute by helping to define the issues. But ultimately, it is the informed citizen who, in a constitutional regime, must be able to understand the issues posed by rapid advances in our scientific knowledge.

While *The Neurotransmitter Revolution* focuses on systems in the brain that are primarily controlled by serotonin and related chemicals, this study is nevertheless intended as a model for a broader range of issues. The life sciences are producing many forms of knowledge whose practical and theoretical effects will require changes in our laws and in our understanding of human nature. Behavior genetics is advancing rapidly, as researchers report genes predisposing their carriers to alcoholism, schizophrenia, and a variety of other behavioral traits and mental illnesses. Cell biology is unraveling the complex processes underlying the formation and development of different living forms. Cognitive neuroscience and neuroanatomy are revealing the characteristics of normal human thought and the influence of developmental abnormalities in such areas as language function, memory, and motor behavior.

Society needs to learn how to discuss these issues. As a first step, this volume addresses a specific set of issues. Much information, reflecting many different approaches, needs to be integrated if new scientific findings are not to undermine our tradition of a system of law in which the rights of the citizen are protected by due process. Each of the contributors has a distinct approach based on a well-developed professional outlook. To understand any of the legal and social problems being generated by the life sciences, one must consider this range of perspectives. None alone provides *the* answer. In a constitutional democracy, what is needed above all is a comprehension of perspectives that have all too often been isolated in our universities and professions. The importance of such an integrative approach is made apparent by the following outline of the substantive arguments we encounter in this volume.

SEROTONIN AND BEHAVIOR (PARTS 2 AND 3)

Part 2 focuses on the neurotransmitter serotonin and the brain systems that are primarily modulated by its activity. In chapter 4, Yuwiler, Brammer, and Yuwiler provide an overview of what is known about the neurochemistry of serotonin. For many nonscientists, this will be a difficult chapter. The reader is nonetheless encouraged to attend to the argument with some care, for—as Yuwiler et al. show—the evidence concerning how much serotonin is active in the brain turns out to be exceedingly difficult to establish. Not only does each of the major neurotransmitters interact with a number of active compounds fulfilling a variety of regulatory functions but the resulting neurotransmitter systems also interact with each other.

Despite this complexity, a picture of the brain and behavior is emerging in which specific behavioral traits or psychological conditions can be associated with relatively low levels of serotonergic function. This volume contains representative work from six research approaches that examine the behavioral effects of serotonergic levels and functioning: Dan J. Stein and Michael Stanley of Columbia Medical School (low serotonin and suicide); Markku Linnoila and his colleagues at the National Institute of Alcohol Abuse and Alcoholism (NIAAA) (studying low serotonin activity and hypoglycemia as factors predictive of recidivism for arson and impulsive homicide); Richard J. and Judith

J. Wurtman of Columbia (seasonal affective disorder, or SAD); Benson E. Ginsburg of the University of Connecticut (individual development and social environment as factors influencing the expression of behaviors for which individuals may have differing genetic predispositions); Michael J. Raleigh and Michael T. McGuire of the University of California at Los Angeles Medical School (high serotonin levels as a correlate of social dominance); and Douglas Madsen of the University of Iowa (individual differences among humans in the behavioral correlates of serotonin).

Stein and Stanley (chapter 5) consider the growing evidence of a link between low serotonergic function and suicide. They survey the sequence of findings that has led to a remarkable shift in the hypotheses used to explain the data. Despite the great complexity of the neurotransmitter system, it is now difficult to dismiss the evidence that suicides or those attempting suicide differ from others in having relatively lower levels of serotonin. But what does it mean?

It was, as Stein and Stanley point out, originally thought that the link between serotonin and suicide was a tendency for individuals with lower levels of the neurotransmitter to be more aggressive than normal. Recently, however, this interpretation has been replaced by the view that the behavioral trait in question is greater *impulsiveness*. For reasons that will become clear in the light of other syndromes associated with poor serotonergic function, this explanation has the advantage of linking serotonin with its effects as a neurotransmitter that integrates the activity of various centers (or "modules") in the central nervous system. Nonetheless, it needs to be stressed that what at first may seem to be a chemical tendency toward aggressiveness seems, in actuality, to be a lack of impulse control, which can be associated with a variety of conditions, many of which have little or nothing to do with aggressiveness.

The work of Linnoila and his colleagues (chapter 6, Serotonin and Violent Behavior) illustrates especially well the extraordinary promise of research on the biochemistry of human behavior, as well as the danger of public misunderstanding of complex scientific findings. Studies in a number of countries have now shown that low levels of serotonin are associated with a wide range of impulsive behaviors, including unpremeditated homicide, arson, and suicide. To explore the mechanisms involved, the NIAAA group has followed a sample of Finnish arsonists and homicidal males in a prospective study in order to ascertain if low serotonergic functioning (as measured by cerebrospinal fluid [CSF] levels of 5-HIAA, the principal metabolite of serotonin) is associated with recidivism.

The findings were quite remarkable, since a combination of low glucose nadir after a glucose tolerance test and low 5-HIAA at the time of original assay correctly classified 84.2% of the subjects in predicting recidivism versus nonrecidivism over the next 36 months. Whereas other behavior and other diagnostic factors do not accurately predict repeated criminal acts among dangerous offenders, it appears that a combination of behavioral and psychobiological variables might hold promise in predicting which individuals are at risk to commit repeated offenses. Of particular interest is the fact that linear discriminant analysis shows no false negatives, suggesting that, even if this method did not predict recidivists with total certainty, it might be used to identify individuals presenting a low risk of repeated offenses and hence likely to benefit from rehabilitation or parole.

Linnoila's work points to three important considerations that are crucial before interpreting the relationship between individual levels of neurotransmitter activity and criminal behavior. First, aggressive behavior may well be a secondary phenomenon in several categories of "violent crime." Despite our tendency to focus on violence as a "cause" of criminality, the mechanisms implicated in this study seem to be associated with *poor impulse control rather than with aggressiveness per se*. For example, low serotonin metabolism is evident in impulsive arson by individuals who do not engage in aggressiveness toward others, as well as in monopolar depressives who attempt or commit suicide.

Second, despite the early tendency to focus on serotonin as a causal factor, low levels of 5-HIAA alone are not sufficient to predict impulsive or antisocial behavior, such as arson or unpremeditated homicide. It was the combination of hypoglycemia—low levels of glucose in response to the glucose tolerance test—and low serotonin turnover that predicted recidivism, not serotonin levels alone. This finding suggests that two rather distinct mechanisms were implicated in those individuals likely to engage in repeated acts of impulsive arson or homicide, a hypothesis that is strengthened by the lack of correlation between glucose nadir after the tolerance test and CSF levels of 5-HIAA or other monoamine metabolites (3-methoxy-4-hydroxyphenlglycol [MHPG] as a measure of norepinephrine turnover and homovanillic acid [HVA] as a correlate of dopaminergic activity).

Finally, and most promising, is the apparent role of alcoholism in the etiology and functioning of those individuals at risk for repeated offenses due to impulsive behavior. Linnoila and his group find that recidivists for arson or impulsive homicide are likely to be type-2 alcoholics (Cloninger, Bohman, & Sigvardsson, 1981), whose impulsive drinking is not associated with guilt. Type-2 alcoholism seems, moreover, to be a sex-linked familial trait, since it is typically found in the father or one of the grandfathers of the recidivists in the sample (Linnoila, DeJong, & Virkkunen, 1989). Why should it be that alcoholism is associated with impulsive asocial behaviors in young males who also exhibit low serotonergic turnover and low glucose tolerance? Can this pattern help explain differences in the effects of those neurotransmitters that are increasingly associated with important social behaviors?

Linnoila suggested an intriguing hypothesis. Although alcohol has the short-term effect of releasing serotonin, over the long term it tends to deplete serotonergic levels. Since it is known that low levels of serotonin are associated with depression, alcoholism may therefore arise as a crude form of "self-medication"; while the immediate disinhibition and release from depression produced by drinking functions as a positive reinforcer, through time alcoholics not only further reduce the rate of serotonin turnover but, at least in individuals with a tendency to hypoglycemia, suffer from particularly acute loss of impulse control.

These considerations have great importance in qualifying the popular reaction to premature reports that serotonergic functioning and perhaps heritable deficiencies in one or more neurochemical process may be associated with criminal behavior. Because low serotonin levels can be treated, the findings hardly suggest biological or genetic determinism. Quite to the contrary, the greatest danger may arise from premature attempts to use scientific findings as a means of behavior control. In particular, not only are measures of a single neurochemical system extremely risky but even complex interactions seem to be sensitive to differences in individual behavior and subject to treatment and control.

Research by Wurtman and Wurtman (chapter 7, Carbohydrates and Depression: Serotonin and Seasonal Affective Disorder) focuses on a very different type of behavior, known as seasonal affective disorder (SAD). For some individuals, the coming of shorter days in the autumn of each year is associated with the onset of depression. These mood changes are, in those affected, often severe and debilitating. Classical explanations, in terms of individual experience or psychological maladjustment, do not seem valid. Now, however, we know that they arise from a combination of low serotonergic activity and deficits in melatonin, another neurochemical system associated with response to light. And the Wurtmans demonstrate with clarity how the interaction of these neurochemical systems explains SAD. In this case, the conventional dichotomy between genes and environment (unfortunately still dominant among nonscientists) is particularly absurd. It is an environmental change that triggers the profound psychological depression of those affected—but the mechanism responsible for the effect is biochemical and probably heritable. The effect of this knowledge, moreover, contradicts the critics of biological explanations of human behavior (e.g., Lewontin, Rose, & Kamin, 1984), since advances in behavior genetics and neurotransmitter research can increase freedom and self-control rather than condemn humans to some form of predetermined fate.

In this case, the research points to specific and easily managed ways of controlling what has hitherto been a devastating psychological condition. By adjusting light levels, many of those afflicted have found that symptoms disappear or are more easily controlled. Hence SAD is a classic example of the way increased knowledge of the biological factors influencing behavior leads to greater individual control. Far from indicating determinism, research into the behavioral effects of neurotransmitter systems can be liberating.

Such paradoxes are also evident in Stanley's research on the presence of low levels of serotonin among suicides (and particularly among those attempting or succeeding in violent suicide). In depression and suicide, the effects of neurotransmitter levels depend in part on the experience of the individual. It is, however, difficult to conduct controlled research on humans to unravel the effects of social experience and individual development on neurotransmitter systems and social behavior. For this reason, the work described by Ginsburg (studying wolves, dogs, coyotes, and mice) and by Raleigh and McGuire (vervet monkeys and other primates) is especially useful.

Part 3 turns to factors influencing the way serotonin works. Ginsburg (chapter 8, Ontogeny, Social Experience, and Serotonergic Functioning) explores experimental studies of social behavior in several species of mammals. Although some individuals in species as diverse as wolves, dogs, mice, and chimpanzees show evidence of "hyperaggressive" behavior that appears to be heritable, the genetic mechanisms associated with behavioral tendencies are normally quite sensitive to individual development and social context.

One example is the defensive-threat behavior of the coyote, an innate response that is not found among dogs. Because coyotes and dogs can mate, even though the two species have different behavioral repertoires, experiments in crossbreeding can illuminate the relationship between genetic and developmental factors in the expression of a specific response. Ginsburg has found that, in some interbred coyote-dog strains, there are "animals carrying the genetic capacity for the coyote defensive-threat behavior along with the genetic system of the dog, [who] . . . spontaneously switch from the dog behavior to the species-typical coyote behavior during or after puberty." The explanation seems to be that the "coyote genes have, in this instance, been brought to active expression by the sex steroids in both sexes," even though the coyote behavior itself is not expressed until "the animal is aggressively challenged."

In this case, an experiment illustrates the combined roles of genetic inheritance, hormonal priming during development, and social context. Since any one of these factors in isolation is not sufficient to produce the particular form of threat behavior, Ginsburg warns that simplistic extrapolations from a single genetic or hormonal factor might be dangerous. This cautionary note is reinforced by evidence from experimental studies on mice, showing that the effects of serotonin on aggression may be due to specific serotonergic receptor sites (rather than to the serotonin levels themselves), as well as to interactions between dopamine and serotonin (rather than to a single neurotransmitter system acting in isolation).

The latter findings are based on studies with two mouse strains that differ dramatically in aggressive behaviors. For one strain (C57BL/6), individual males reared in isolation and introduced to a dyadic situation are far more aggressive than the other (BABL/c). Moreover, each mouse strain tends to show different aggressive behaviors, with "attack" and "chase" accounting for 90% of agonistic behavior in C57BL/6-strain males, but only 28% in the BABL/c strain (in which "wrestle" and "tail rattle" account for 70% of the agonistic actions).

Since males of the two strains thus appear to have different innate behavioral repertoires, crossbreeding the two provides an elegant way of studying the interaction of experience, individual development, and social context in the expression of aggressive behaviors. Experiments using amphetamines that act as a dopamine agonist produce marked effects on aggressive behavior, but the response is diametrically opposed in each of the two mouse strains: "where the effect on one genotype is to increase aggression, the effect on another is to decrease it." As a result, a statistical prediction of neurotransmitter influence on a broadly defined category of behavior like aggressiveness is "simply a statement of odds that cannot serve as a predictor for any specific instance."

This need for caution is reinforced by a series of studies conducted by Michael J. Raleigh and Michael T. McGuire at the UCLA Neuropsychiatric Institute (chapter 9). Focusing on vervet monkeys, Raleigh and McGuire have found that individual experience and social environment have an important effect on the expression of aggressive behaviors associated with serotonergic functioning. By using a variety of drugs, this group found that, in vervets as in humans, reduction of serotonin turnover is associated with increases in aggressive behavior—but that the relationship is far from simple.

Over a number of years, Raleigh and McGuire have found that, even though dominant and subordinate vervets do not differ physiologically, the dominant male in a vervet group has higher serotonin levels and lower rates of aggressive behavior than subordinates. These effects seem to be mediated by social interaction, since it is the sight of submissive behavior by others in the group that triggers the neurohormonal changes leading to enhanced serotonin turnover and dominant social behaviors. Similar findings in other species confirm that social experience and group structure may be extremely important in the functional relationship between serotonin and behavior.

Among the experimental findings of particular relevance to the human data presented by Linnoila is the effect of social rank on the target of aggressive responses within the group. In a vervet band, some individuals, such as the young, are not typically "appropriate" as targets of aggressive or threatening behavior. When serotonergic functioning is experimentally reduced, subordinate vervets—but not dominants—are more likely to attack socially "inappropriate" individuals. Subordinate vervets whose serotonin is artificially lowered thus have what might be called "difficulties in impulse control," perhaps comparable to Linnoila's findings among recidivist prisoners. Because similar treatments of a dominant male do not have this effect, however, clearly the functional consequence of serotonin on behavior depends on both the individual's rank and the social situation.

One might well wonder how those effects due to status and social situation that have been observed in nonhuman primates relate to humans. Douglas Madsen's work on the role of serotonin in humans helps provide an answer. Madsen's previous studies had shown that in our own species, as among vervets, leaders tend to have higher serotonin levels than do followers (Madsen, 1985). In chapter 10, he presents evidence to show that the behavioral effects of serotonin are complex. Individuals with higher-than-average levels of serotonin are not always those with the most aggressive responses; for some, the effect is just the opposite. Once again, the simplistic view that a single chemical can "cause" behavior is contradicted by the evidence.

THE CHALLENGE TO LEGAL CONCEPTS (PART 4)

Part 4 turns to the legal and social implications of the scientific findings presented in this book. Two contrasting perspectives illuminate the problems created by the "neurotransmitter revolution." A vigorous challenge to prevailing attitudes is offered by C. Ray Jeffery (chapter 11, The Brain, the Law, and the Medicalization of Crime), who makes the case for the "medicalization" of crime and an abandonment of the mens rea concept as both unscientific and impractical. In contrast, Michael H. Shapiro (chapter 12, Law, Culpability, and the Neural Sciences) argues that new research does not require abandoning the traditional concept of legal responsibility.

For Jeffery, our approach to criminal justice combines an irrational model based on revenge and retribution with a rational model based on deterrence. The failure of this approach to confront crime effectively has combined with the emergence of new, biomedical sciences of behavior to make possible a medicalization of crime. More specifically, discoveries concerning the role of amino acids in the brain make it possible to move from ineffective punishment to coherently defined prevention and treatment of individuals at risk of violence.

For Jeffery, the inability of our prisons to function either as rehabilitative environments or as effective deterrents increases the relevance of therapeutic alternatives. If attention is shifted to "crime prevention," a prisoner's "right to treatment" might become more salient than such issues as "informed consent," which now often dominate discussions of medical treatment of prisoners. While mentalist approaches to crime and mental illness persist, discoveries in the physical and chemical bases of normal, as well as abnormal, behavior have spread rapidly. Given the likely impact of findings based on new scientific theories and techniques, Jeffery argues in favor of moving from punishment to prevention and treatment for crime.

Michael H. Shapiro's argument in chapter 12 (Law, Culpability and the Neural Sciences) begins from the premise that the legal system has typically assumed that human behavior is "caused." On this premise, new scientific findings provide particular details concerning causation but need hardly be grounds for radical changes in the law. After surveying the varied theories of freedom and responsibility, Shapiro suggests that knowledge of the "physical substrates" of behavior would not be likely to change the concept of culpability unless it could be shown that an individual lost preexisting capacities for reason and deliberation. From this perspective, although nothing in new research compels a move toward a "therapeutic" model of jurisprudence, future biological discoveries could lead us to modify or "reassemble" our notions of guilt, confidentiality, and other basic norms.

While the positions of Shapiro and Jeffery are not entirely inconsistent, they do set forth two rather different ways of responding to the kinds of data emerging from biobehavioral studies of the effects of serotonin and other neurotransmitters. In the discussion at the "Serotonin, Social Behavior and the Law" conference, other participants pointed to the risks of seeking to define a "perfect" legal "solution" to the ambiguities of apparently contradictory doctrines. As E. Donald Elliott of Yale Law School has noted, for example, attempts to change the law by means of frontal assaults have typically had effects quite opposite from those intended (cf. Elliott, 1985). Hence, before deciding whether or not to modify such a fundamental concept as responsibility, it is important to consider in more detail the scientific evidence in question.

IMPLICATIONS FOR LAW AND POLICY (PART 5)

The discovery that some individuals may have a genetic predisposition to impulsive violence or deviant behaviors, which are associated with neurotransmitters like serotonin, could have far-reaching effects, particularly if the many qualifications noted above are not considered when this research is introduced into either criminal or civil law. To gain further perspective on the appropriate way to integrate scientific knowledge into the legal system, part 5 begins with a chapter on the way our legal systems evolves.

More often than not, the law reflects accidents and paradoxical consequences, which suggests that it is dangerous to propose legislation or public policy on the assumption that intended results predict likely outcomes. William H. Rodgers, Jr. (chapter 13, The Lesson of the Owl and the Crows) provides evidence of the general point by analyzing the role of deception in the evolution of the environmental statutes. Models in game theory that have been used to describe animal behavior can be applied equally well to legal decisions and doctrines, such as those in the area of environmental law. In predator-prey relations, deception is often highly advantageous—at least until counterdeception evolves to counteract the fake.

Confronted with predation from an owl, for example, crows "developed a strategy of wandering into easy range 'pretending' to be wholly unaware of the presence of the owl, only to sidestep the futile strikes with disdain and ease." In much the same way, Rodgers shows, the development of environmental statutes has been marked by strategies of deception, using such techniques as procedural entitlements, ambiguity, delegation of authority, postponements, and self-nullifying legislative provisions.

If the law is a jerry-built system of "bricolage" in which the legislative process is typically characterized by deception and counterdeception, simple adjustments to new scientific evidence concerning social behavior are unlikely to be productive. On the contrary, it will often be most valuable to seek indirect means of integrating new information into the legal system. In no area is this more likely to be sound advice than in the biological correlates of criminal behavior, if only because of the enormous passions released by fears of violence and anger at seemingly ineffective judicial procedures.

The conclusions that can be derived from research on serotonergic function thus touch many issues and challenge conventional concepts of the relationship between nature and nurture. The scientific evidence demonstrates that the loss of impulse control associated with neurochemical imbalances depends greatly on individual experience or social situation and can produce diverse forms of deviant behavior (e.g., alcoholism, arson, suicide, homicide). Insofar as knowledge of the biological causation of behavior is being advanced, we learn more about how to *control* outcomes; rather than genetic determinism or biological reductionism, the complexities of the serotonergic system point to possibilities of prevention and treatment that require caution if they are not to undermine our traditions of legal procedure.

In discussing this extraordinary range of research linking the life sciences to the study of law, the most interesting questions concern how to minimize dangerous error and misunderstanding. Many defense lawyers have already sought to benefit from the apparent correlation between low serotonergic function (accompanied by hypoglycemia) and violent social behavior. If the public were to be convinced of such findings, the consequence might well be a demand for genetic screening, possibly combined with extensive economic, social, and legal discrimination against individuals found to carry supposedly defective genes.

In chapter 14 (Low Serotonin Function and the Therapeutic Power of the Criminal Law), David B. Wexler turns the concept of "medicalization" of crime on its head by considering the criminal law as *itself* a therapeutic system. Laws are not merely a means of punishing individuals who have been tried for and convicted of crimes. The legal system as a whole needs to be seen, at a broader level, as a means by which social norms are transmitted and the behavior of citizens shaped. As Wexler shows, the problem of criminal behavior linked to neurotransmitters like serotonin needs to be assessed in this context.

From another perspective, the research on the behavioral effects of serotonergic function concerns an unexpected set of legal and social issues. As Steven Goldberg (chapter 15, Science Spending and Serotonin) argues, biological knowledge obviously modifies the legal process and hence can play a central role in changing the efficacy of governmental programs and laws. If so, the level and distribution of federally funded scientific research on neurotransmitters and behavior may be a critical question of public policy. Among the implications of the findings presented in this volume, it is not the least important to realize how political decisions can influence the extent to which future generations will understand how citizens respond to political decisions.

A major suggestion arising from the conference out of which this volume grew concerned the relationship between serotonergic functioning and legal procedures that are not associated with criminal law. In the final chapter (Conclusions for Public Policy: Early Intervention, Special Education, and the Law), I propose that, insofar as some individuals may be genetically at risk for low serotonin turnover and hypoglycemia, this condition be considered an entitlement for special education rather than a matter of reduced criminal responsibility. In effect, individuals with the recidivist profile outlined by Linnoila have typically exhibited great difficulties with impulse control in elementary school; such behavioral problems are often associated with poor educational results, marginality within the school, and persistent social failure.

Whereas evidence of serotonergic functioning or hypoglycemia poses thorny problems when considered in the context of our doctrines of legal responsibility and guilt, the same information might be less likely to give rise to abuse as an entitlement to educational benefits for the young child. On the one hand, the introduction of neurochemical and genetic information into the legal procedure is likely to produce diverse and passionate emotional responses on the part of judges and juries; in such a context, the social and developmental factors involved in the expression of genetic propensities are unlikely to be weighed carefully or accurately. On the other hand, a gene for low serotonin turnover might very well be compared to the genetic substrate involved in some forms of dyslexia.

Scientific information about hormonal and neurotransmitter functioning might therefore be most appropriately viewed as providing parents with an *entitlement* to special educational services for children rather than as a means of diminishing the responsibility of adults. Not only would such a procedure avoid the risk of genetic screening as a mode of criminal prevention but it would address more directly the finding that the violent behaviors of this type of recidivist may well be primarily associated with impulse control and only secondarily with aggressiveness as such.

Among the reasons noted in favor of this approach is the risk of genetic engineering or of stigmatizing those individuals who might carry genes linked to low serotonin turnover and hypoglycemia. Such genes have presumably been maintained in the human gene pool for some reason: it would seem, for example, that those males affected are more likely to engage in the risk-taking activities associated with altruistic bravery in warfare. Ill-conceived attempts to "cure" crime by genetic engineering might well backfire.

In both nature and law, evolution is a complex process. As Elliott likes to put it (cf. Elliott, 1985), legal doctrine is often like a panda's thumb, resulting from a kind of bricolage as old structures take on new functions. While the public seeks certainty when learning of scientific research into human behavior, the relationships among genetics, neurotransmitters, and social behavior are complex and sometimes unpredictable. Despite the wide variety of disciplines represented, the participants at the conference from which this volume emerged were agreed most emphatically on the dangers of assuming that simple solutions might be found in responding to future developments in the neurochemistry of behavior.

REFERENCES

Cloninger, C. Robert, Bohman, M., and Sigvardsson, S. 1981. Inheritance of Alcohol Abuse: Crossfostering Analysis of Alcoholic Men. *Archives of General Psychiatry* 38:861–68.

Elliott, E. Donald. 1985. The Evolutionary Tradition in Jurisprudence. *Columbia Law Review* 85:38–94.

Lewontin, Richard, Rose, Steven, and Kamin, Leon. 1984. *Not in Our Genes.* New York: Pantheon.

Linnoila, Markku, DeJong, Judith, and Virkkunen, Matti. 1989. Family History of Alcoholism in Violent Offenders and Impulsive Fire Setters. *Archives of General Psychiatry* 46:613–16.

Madsen, Douglas. 1985. A Biological Property Relating to Power-Seeking in Humans. *American Political Science Review* 79:448–57.

Part 2

Serotonin and Behavior

Increasing understanding of the role neurotransmitters play in human social behavior is illustrated by the varied functions of serotonin.

4

The Basics of Serotonin Neurochemistry

Arthur Yuwiler,
Gary L. Brammer,
and K. C. Yuwiler

Abstract: A technical survey of research on neurochemistry and behavior shows the complexity of the functions and effects of serotonin and its associated neurotransmitters. These findings demonstrate the difficulty of measuring the level of serotonin and the dangers of simplistic conclusions concerning its role in causing behavior. Because this chapter is difficult, readers without scientific training may prefer to begin with the remainder of part 2 and come back to it later for an understanding of technical issues, which are essential for those who address the policy implications of serotonin and human behavior.

Legal interest in biological concomitants of behaviors stems from the desire to apportion deterministic and volitional contributions to human actions and, by so doing, help establish legal culpability. The suggestion that serotonin may serve such a function comes from several sources. Not only is the serotonin system in the brain linked to emotionality but, as reviewed elsewhere in this volume, the concentration of serotonin and its metabolites in accessible body fluids has been reported to be abnormal in various socially significant states like depression, aggression, and anxiety. Its chemical relationship to the weak hallucinogens bufotenine from the toad, psilocybin from the sacred mushroom of the Aztec, N,N-dimethyltryptamine, as well as LSD-25, provides additional impetus for linking serotonin with aberrant behavior. However, before deciding if concentrations of serotonin and its catabolites in tissue can be used as a kind of medical phrenology or if they provide a basis for behavioral prognosis, it may be useful to review some aspects of neurobiology, the control of serotonin metabolism, and the relationship between the brain and peripheral serotonin.

Nerves communicate largely by chemical signals. Currently there are some dozen small molecules and 50 or more peptides thought to be neurotransmitters; these are defined as compounds released from one neuron that then bind to specific receptors on another, adjacent neuron to produce a physiological change. Besides serotonin, the "classic" transmitters include norepinephrine, epinephrine, dopamine, acetylcholine, glutamic acid, aspartic acid, glycine, gamma-aminobutyric acid (GABA), histamine, and adenosine. Representative peptide transmitters are vasoactive intestinal polypeptide, cholecystokinin, leucine enkephalin, methionine enkephalin, substance P, peptide Y, neurotensin, somatostatin, thyrotropin-releasing hormone (TRH), bombesin, luteinizing hormone-releasing hormone, adrenocorticotropic hormone, beta endorphin, vasopressin, oxytocin, and carnosine.

Transmitters are released from the presynaptic neuron to cross a short synaptic space and bind to a receptor on the postsynaptic target neuron. This stimulates the postsynaptic neuron and initiates a set of reactions within the cell or in its membrane that affects cellular metabolism and release of transmitters to a subsequent neuron, until the sequence is halted by other neural processes or terminates in a completed action. Each neuron releases at least one specific and characteristic transmitter. Thus, there are cholinergic neurons, releasing acetylcholine; serotonergic neurons, releasing serotonin; and so on. Some also release a second, a cotransmitter, as well. Most serotonergic neurons, for example, release only serotonin, but some also release the peptide substance P, others TRH, and still others opioidlike enkephalins. Substance P, TRH, and enkephalins are also neurotransmitters in their own right, and the neurons they cohabit with serotonin could as easily be designated substance P, TRH, or enkephalin neurons with serotonin as a cotransmitter. The significance of corelease is generally unknown, although in some instances one cotransmitter can modulate the effects of the other.

While receptors are specific for a particular transmitter, they can be divided into subtypes on the basis of differential sensitivity to various pharmacological agents and on the basis of the metabolic effects initiated when the receptor is stimulated. Receptors are complex proteins, and receptor subtypes differ both chemically and structurally, although they do have many structural features in common: indeed, structural similarities between some receptors are so marked that it has been suggested that one form mutated from another. Most transmitters act on several receptor subtypes, and the different subtypes often have distinctive anatomical distributions. Subtype identification is an active field of research at this writing, and new receptor subtypes are proclaimed almost daily. Each neuron, of which there are some 10 billion in the brain, may receive chemical signals from as many as 10,000 sources or from as few as one. Each source may release a different transmitter, which may act on one of several receptor subtypes. Some may be stimulatory to the target neuron and some inhibitory. That is, some may increase (or decrease) neuronal excitability by lowering (or raising) its membrane potential. The actual state of the target neuron reflects an integration of these many influences according to their type, strength, and synaptic location.

Thus, the effect of any particular transmitter on a neuron is at least tempered by the status of other transmitters affecting that neuron.

The transmitter of primary interest here, serotonin, or 5-hydroxytryptamine (5-HT), is derived from the amino acid tryptophan. This is oxidized to 5-hydroxytryptophan (5-HTP), and an acid group (carboxyl) is then removed from 5-HTP to yield 5-HT. Then 5-HT is degraded in the brain by oxidation to 5-hydroxyindoleacetaldehyde, which is then further oxidized to 5-hydroxyindoleacetic acid (5-HIAA). This exceedingly simple metabolic scheme is operationally complex (figure 4.1).

The first and rate-limiting step in this pathway is the combining of tryptophan with one atom of molecular oxygen (by the enzyme tryptophan hydroxylase) to produce 5-hydroxytryptophan (5-HTP) and an atom of reactive oxygen. To carry out this oxidation, an enzyme cofactor, tetrahydrobiopterine, is needed, which undergoes reversible oxidation and reduction, coupling to another enzyme and another electron recipi-

Schematic Serotonergic Neuron Illustrating Synthesis, Release, And Metabolism Of Serotonin, As Well as Pre- And Postsynaptic Binding Sites

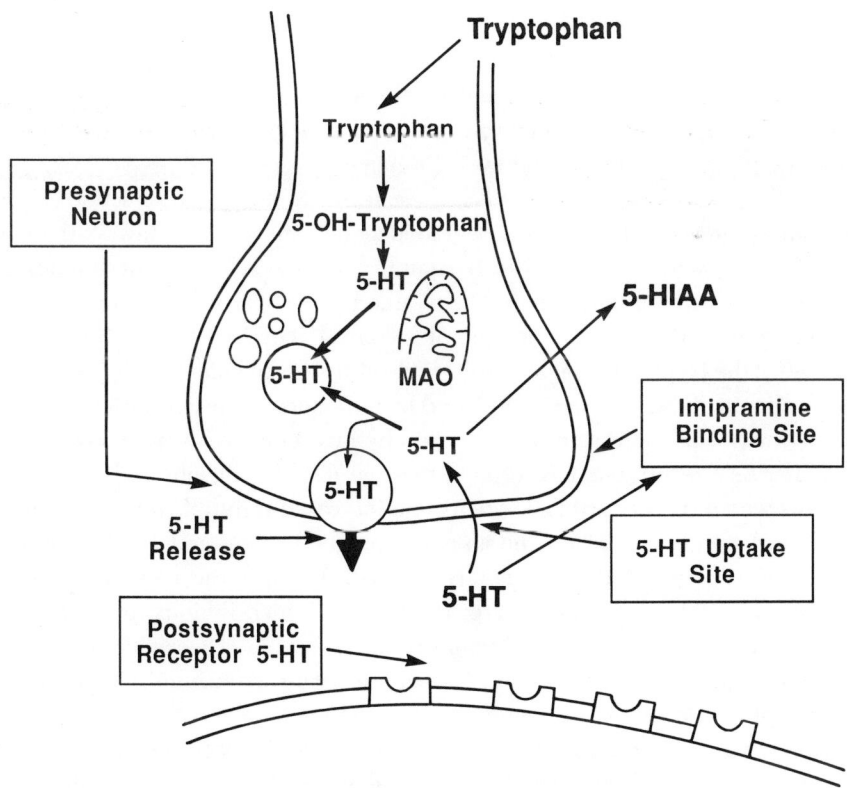

Figure 4.1. Schematic serotonergic neuron illustrating synthesis, release, and metabolism of serotonin, as well as pre- and postsynaptic binding sites.

ent. Although tryptophan hydroxylase is the rate-controlling enzyme, the rate of reaction depends upon both enzyme activity and substrate concentration. This is unusual at the first step in a metabolic sequence. More commonly, the rate of this step is controlled by enzyme activity alone, which in turn is regulated by feedback inhibition due to some product generated along the metabolic path. The dependence of the rate of tryptophan hydroxylation on both enzyme activity and tryptophan concentration results from the fact that the enzyme is not normally saturated with substrate. In reactions that are controlled by the enzyme activity *only*, the enzyme is binding as much substrate as possible at all activity levels. Additionally, the rate of 5-HTP formation is affected by other substrates in the reaction (oxygen and the cofactor tetrahydrobiopterine) and by the activity of the system regenerating reduced cofactor. The decarboxylase step occurs almost immediately to produce serotonin. Thus, tryptophan loading, both in animals and in man, increases tissue serotonin. Predicting brain serotonin from tryptophan intake, however, is complicated by tryptophan transport and metabolism.

While serotonin is formed within the body, its essential precursor, the amino acid tryptophan, is not, creating the interesting situation of a central nervous system transmitter being dependent upon the exogenous supply of its precursor. Serotonin is not alone in this feature, however: the precursor to the catecholamines also must be supplied from the diet. Tryptophan not only must be supplied in the diet but is not overly abundant in most foods. Besides its role as precursor to serotonin, tryptophan is essential for protein formation. It is a constituent of protein, but it also stimulates protein synthesis. Moreover, tryptophan affects an important step in energy production and thereby has an important role in diverting metabolism from protein synthesis to energy production as external events demand.

Only about 1% of dietary tryptophan is converted to serotonin; a fraction is incorporated into protein, and the bulk is degraded along a metabolic pathway, the kynurenine pathway, which leads to the B vitamin nicotinic acid in most animals but to compounds of little or unknown biologic value in man. Tryptophan breakdown along the kynurenine pathway is under complex control. The first step is the irreversible cleavage of the tryptophan molecule carried out by the enzyme tryptophan pyrrolase. The formation of this enzyme is increased by hormones of the adrenal cortex released during stress; this increase can be blocked by high glucose levels and/or by growth hormone. As a result, animals respond to some, but not all, physical and emotional stressors with a rise in pyrrolase activity and increased tryptophan breakdown. Their response depends in part on the animal's metabolic state and whether hormones other than adrenocortical hormones are released. Further, the amount of tryptophan pyrrolase is affected by its substrate, tryptophan, which inhibits breakdown of the enzyme and thereby promotes its accumulation. A combined increased rate of enzyme formation—due to corticoids—and a decreased rate of enzyme breakdown—due to tryptophan—leads to "superinduction" of the enzyme and a superdestruction of available tryptophan stores. The sequence of a stress-induced increase in pyrrolase activity leading to increased tryptophan metabolism, decreased serum blood tryptophan, and decreased brain serotonin seems to occur with some, but not all, stressors presumably because of associated hormonal and metabolic changes.

In most animals, this pathway forms three compounds, each of which can affect the brain: nicotinic acid, kynurenic acid, and quinolinic acid. As indicated above, nicotinic acid is a B vitamin that prevents pellagra. Quinolinic acid appears to be transported into the brain, where it has potent excitatory effects on a particular type of excitatory amino acid receptor, the N-methyl-D-aspartate (NMDA) receptors. On the other hand, kynurenic acid is a broad spectrum antagonist at excitatory amino acid receptors (where it blocks the action of quinolinic acid) and at the glycine receptor linked to the NMDA receptor. Kynurenic acid can be transported into the brain from blood and can be made within the brain from kynurenine.

To be converted to serotonin in nerve cells, tryptophan must get to those cells. This too is complex. Tryptophan itself is a large molecule and relatively insoluble in aqueous solution, so that it is largely carried in blood by being bound to specific sites on serum albumin. Only about 10% exists in free solution. Tryptophan's binding to albumin is sensitive to blood pH and factors that make for acidosis or alkalosis. Tryptophan can also be displaced from its albumin binding sites by drugs like aspirin or by endogenous compounds, especially free fatty acids released from fat stores by adrenaline during stress. Since albumin is too big to penetrate brain capillaries, only free tryptophan and tryptophan released from its binding sites is transported. Indeed most, but not all, albumin-bound tryptophan is stripped from its binding sites during passage through the brain's capillaries, so that total blood tryptophan levels, as well as free tryptophan, affect quantitative tryptophan transport. While the free-to-bound ratio influences tryptophan uptake into tissues, the relative concentration of tryptophan to other neutral amino acids is an even more important regulator of quantitative tryptophan uptake into the brain because blood capillaries in the brain are built to exclude materials from entering unless they are very soluble in fats that make up cell membranes and in water that makes up the cell substance, or unless they are carried through the capillary walls by specific transport carrier systems.

There is one carrier system for all the neutral amino acids, including tryptophan; and all neutral amino acids compete for sites on the carrier with an effectiveness that depends on concentration and affinity for the carrier site. Tryptophan has a relatively high affinity; other neutral amino acids with high affinity are leucine, isoleucine, phenylalanine, and tyrosine. Interestingly, the mental retardation in genetic diseases like phenylketonuria and branched-chain ketonuria may result from excessive phenylalanine or leucine inhibition of the transport of other neutral amino acids. Thereby, protein synthesis is impaired during the early period of neural connectivity. Of course, each of these competing amino acids is also under complex dietary, hormonal, and metabolic control.

In sum, the acquisition, metabolism, and transport of tryptophan all influence serotonin biosynthesis and add a layer of complication to the first simple step in its synthesis. The subsequent step, the decarboxylation of 5-HTP to serotonin, is rapidly carried out by a ubiquitous pyridoxal-phosphate-requiring enzyme, aromatic amino acid decarboxylase, which is common to serotonergic, dopaminergic, noradrenergic, and adrenergic neurons.

Serotonin synthesis, then, consists of only two enzymatic steps. Serotonin degradation is even simpler. It is oxidized by monoamine oxidase to 5-hydroxyindoleacetaldehyde, which is oxidized further to the acid 5-HIAA (5-hydroxyindoleacetic acid) or reduced to the alcohol 5-hydroxytryptophol. In the brain the product is 5-HIAA.

The enzyme monoamine oxidase is localized on the outer membrane of the mitochondria, and two different forms exist, MAO A and MAO B, which differ in substrate specificity, drug sensitivity, and structure. There are currently conflicting pharmacological and immunohistochemical data on which form exists in serotonin neurons and degrades serotonin in vivo. While these issues are of importance in understanding serotonin metabolism, the arguments will not be detailed here. Suffice it to say that serotonin is probably degraded largely by MAO A, and the biochemical and pharmacological data have generally been interpreted as indicating that serotonin is degraded within the serotonergic neuron.

Thus, a controlled fraction of dietary tryptophan is delivered to the brain, where it is converted to serotonin. Thereafter, this newly formed serotonin either is degraded by MAO largely to 5-HIAA or is transferred to storage vesicles in the cell. Stimulation of the neuron leads to a change in internal calcium followed by fusion of the vesicles with the cell wall and release of serotonin into the synaptic space, where it stimulates postsynaptic receptors. Serotonin is largely removed from the synaptic space by a reuptake process back into the serotonin neuron, where it is either degraded by MAO or again stored in synaptic vesicles. Intrasynaptic concentration, then, can be affected by the rate of release, by how much is released at each firing, by the efficiency of reuptake, and by the rate of serotonin degradation.

It is the presence of tryptophan hydroxylase that determines whether a neuron is a serotonergic neuron. The other enzymes in the synthetic pathway are common to catecholaminergic neurons as well. In point of fact, serotonergic neurons make up only about 0.01% of the total number of nerve cells in the brain; but from localized cell clusters in the old part of the brain, they send thin, ubiquitous fibers throughout the whole of the brain and exert broad control over neural events. Serotonin cell bodies are highly localized to nine cell clusters (raphe nuclei) in the base of the brain. From these nuclei arise three major nerve-fiber pathways, called the mesostriatal, mesolimbic, and medial ascending pathways, which ascend into and innervate most of the areas of the brain involved in higher functions and emotion. The cells in different nuclei are more heavily represented in some of these tracts than in others, and the nerves project to different parts of the brain. One pathway, the mesostriatal tract, terminates in the motor-modulation area of the brain, the basal ganglia. The mesolimbic, or ventral ascending, pathway enters the hypothalamus and other areas of the limbic system, or emotional part of the brain. The medial pathway, with heavy contributions from the medial raphe nucleus, projects mainly to the substantia nigra, the site of dopamine cell bodies; the projection fields of these dopaminergic cells is the caudate-putamen complex. Serotonergic cells from the dorsal raphe make a heavy contribution to the caudate-putamen complex. That is, serotonin appears to regulate both the major input to the system and the activity within the system. There are also two main descending projections from posterior raphe nuclei. A bulbospinal pathway innervates the spinal cord and is linked to endogenous analgesia. The second pathway goes from the dorsal raphe to the locus ceruleus, the main site of cell bodies of noradrenergic neurons. There are also several more minor pathways.

Serotonergic neurons, then, not only innervate large areas of the brain but also act on noradrenergic and dopaminergic cell bodies. In reciprocal manner, noradrenergic neurons innervate serotonergic cell bodies especially in the dorsal raphe. Indeed, 67% of the nerve-ending terminals of the noradrenergic neurons in this nucleus make contact with serotonergic neurons. Epinephrine-containing neurons from further back in the brain stem also appear to innervate some serotonergic neurons, as do dopaminergic fibers from the substantia nigra and the ventral tegmentum. In addition, there are some dopamine-containing cells within the dorsal raphe nucleus that may interact locally with serotonergic cell bodies. Peptide transmitters are also found in various raphe nuclei. Some of these peptide transmitters likely act on serotonin neurons. These include vasoactive intestinal polypeptide, vasopressin, luteinizing hormone-releasing hormone, and oxytocin.

Serotonin released from nerves exerts its effects by actions on a receptor. There are currently seven different serotonin receptor subtypes proposed, although not all have been established. These seven subtypes are designated 5-HT1A, 5-HT1B, 5-HT1C, 5-HT1D, 5-HT2A, 5-HT2B, and 5-HT3. (A 5-HT4 and a 5-HT5 receptor subtype have also been claimed.) The 5-HT1B receptor is an autoreceptor at the end of the nerve terminal that regulates serotonin release. It is found in the rodent but not in man, where the 5-HT1D receptor takes its place. Stimulation of the 5-HT1A, 5-HT1B, or 5-HT1D receptors leads to inhibition of one type of intracellular process, the formation of adenylate cyclase, which acts as a second messenger in altering cellular metabolism. Stimulation of 5-HT1C and 5-HT2B receptors leads to activation of a different intracellular processes, the turnover of phosphoinositol phospholipids and the formation of diacylglyceride; these processes influence other aspects of cellular metabolism. The precise intracellular consequences of stimulating the other 5-HT receptor subtypes has not yet been established.* These various receptor subtypes have distinct anatomical distributions in the brain, so that the consequences of a serotonin discharge on local brain activity varies by region. Drugs acting through the serotonin system also have different actions in different parts of the brain in accordance with the distribution of receptor subtypes. In addition, species differ in their cadre of receptors.

On the assumption that the brain directs behavior, the status of the brain's serotonin system is most relevant to the legal question of serotonin and accountability. However, the bulk of serotonin in the body is formed in the gut, where serotonergic neurons carry out the very important business of peristalsis, rather than in the brain, where they may influence the more secondary process of behavior. Moreover, brain tissue is not sampled in a living man; rather, the status of the central serotonin system is inferred from the content of serotonin and its metabolites in blood, in urine, or, occasionally, in cerebrospinal fluid (CSF). The serotonin in each of these tissues derives from different sources, and each represents a particular physiological compartment.

Urine can be to biochemists what middens are to archaeologists. It is the great garbage heap of the soluble products of dietary intake and bodily metabolism collected over time. Urine composition can provide important information about the metabolic state of the body as a whole, roughly over the time period between voidings. Urine is an extraordinarily complex mixture in which a particular compound is represented not only by itself but also in the form of its catabolites. In the case of serotonin, these products come from the central nervous system, the gut, and a variety of peripheral tissues producing or metabolizing serotonin. These include tissues as diverse as the pineal gland and the kidney. Serotonin in urine is represented largely by 5-HIAA, although there is some free serotonin, some 5-hydroxytryptophol, some methoxylated serotonin derivatives, and some degradation products formed after cleavage of the indole ring by brain and liver oxygenases. The major portion of urinary 5-HIAA derives from the breakdown of serotonin formed in the enterochromaffin cells in the intestinal wall. The sum of these metabolites reflects total serotonin metabolism in the body.

In contrast to urine, blood is a sample of the metabolic traffic between tissues at the time the blood sample is drawn. Serotonin in blood is made by specialized cells in the intestine and then released into the circulation, where it is taken up into storage vesicles in blood platelets by an active, sodium-dependent uptake mechanism. Blood serotonin cannot contribute to brain serotonin and brain serotonin does not affect blood serotonin except as the brain might influence the activity of the gut or the release of hormones that might influence intestinal serotonin production or release or serotonin uptake into the platelet. In such conditions as carcinoid syndrome, serotonin levels in the two compartments move quite independently; in such conditions as phenylketonuria, serotonin concentrations are depleted in both compartments, although the depletions occur by different mechanisms. On the other hand, brain and blood serotonin levels are affected in the same manner by precursors and a number of drugs and presumably by endogenous alterations in metabolic key enzymes, serotonin uptake, transport, or storage. Although the content of many compounds in blood changes moment by moment, blood serotonin levels are remarkably stable through time within an individual, even though individuals vary widely in their blood serotonin content.

The CSF is sort of the urine of the adjacent central nervous system. Like urine, it lacks anatomical information and provides even less information than urine on qualitative and quantitative metabolism since many of the brain's metabolites, including those that appear in the CSF, are also pumped directly into the blood. Serotonin in CSF is mostly represented by 5-HIAA. Occasionally, cisternal CSF can be obtained, and its 5-HIAA content reflects that of the brain. More commonly, CSF samples are taken from the lumbar region of the spinal column below the end of the spinal cord. While the brain contributes to 5-HIAA in lumbar CSF, a major portion arises from the spinal cord. Since it takes about six hours for substances to travel from the brain to the lumbar region, the brain's contributions to lumbar 5-HIAA are both diluted and delayed. CSF 5-HIAA at best represents serotonin metabolism, not necessarily serotonergic transmission. While it is true that serotonin entering the synaptic space and stimulating the postsynaptic neuron is largely recovered by reuptake and oxidized to 5-HIAA, the bulk of serotonin may be catabolized intracellularly without ever being released into the synaptic space. Since 5-HIAA can be cleared from the brain through the blood, CSF 5-HIAA represents only a fraction of the 5-HIAA formed in the brain. It is not clear whether even cisternal CSF reflects the whole of the brain's serotonin metabolism. Further, CSF 5-HIAA concentrations correlate with the concentration of the dopamine-oxidation product, homovanillic acid (HVA), coming largely from the motor-modulation center, the caudate nucleus. The correlation has been variously explained as due to saturation of the carrier system transporting organic acids out of the CSF, to dopaminergic control of serotonin turnover, or to serotonin control of dopamine turnover. The real explanation has yet to emerge, but these problems cloud our understanding of the exact relationship between CSF 5-HIAA and whole-brain serotonergic transmission.

This background material suggests that considerable caution is needed in assigning legal significance to tissue concentrations of serotonin or its metabolites. The most critical question, of course, is specificity. In a sense, the concentration of an endogenous compound is a kind of circumstantial evidence, and like all circumstantial evidence, its strength depends on strength of association. The correlation between metabolite concentration and its proposed behavioral concomitant must be nearly unity to be useful. A mere statistically significant difference in substance concentration between populations is insufficient, particularly when there are statistically significant relationships between that substance and other behavioral conditions as well. After all, a single variable can only go up, down, or stay the same; and, as indicated above, there are many factors that can move it in one direction or the other. While blood serotonin may be elevated in the dominant male vervet monkey, it is also elevated in about 30% of autistic children and children with nonspecific mental retardation, but it is lower in retarded children with phenyketonuria and in children with Down's syndrome. The mechanism elevating the blood serotonin level in the monkey presumably differs from that elevating it in the autistic child, while the mechanism for the decreased blood serotonin in the untreated phenylketonuric child differs from that elevating the serotonin concentration in the child with idiopathic retardation. Further, the relationship between brain and blood serotonin concentrations varies in accordance with the mechanism: both are diminished in phenylketonuria, but only blood levels change in carcinoid syndrome. A similar set of comparisons can be made for CSF 5-HIAA. Mechanism is generally more likely to provide diagnostic differentiation than concentration alone, although concentration alone, if the correlate with behavior is strong enough, may be useful even without knowledge of mechanism.

It should also be apparent that concentrations are controlled differently in different tissues and that an abnormality in one tissue may or may not be related to that in another. Additionally, an abnormality in serotonin metabolites in peripheral tissue may not necessarily indicate an abnormality in the brain's serotonin metabolism or turnover. Not only may the two be controlled differently but serotonergic neurons affect other transmitter systems and other transmitter systems affect serotonergic neurons, so that the normal expression of a normal serotonin neuron may be perverted by an abnormality in its postsynaptic nonserotonergic partner; or, conversely, an abnormality in a serotonin neuron's presynaptic input may lead to an abnormal expression by a normal serotonin neuron.

And, of course, individuals vary greatly, and most of our understanding of biology is derived from the biology of groups, not the biology of individuals. Every variable has a distribution, whether Gaussian or not, and some values are far from the mean.

Of the multitude of metabolic systems that comprise the human machine, some are skewed from the mean, and our personal adaptations to that skewing make us the unique creatures we are. Every clinician has stories of abnormal responses to treatments and of seemingly normal functions despite grossly abnormal values on some biological variables. Within limits, one person's abnormality may be another's normal state. This is what keeps medicine an art, as well as a science, and such medical uncertainties lead to legal ones as well.

Finally, for those of us who assert that neurochemistry, neurophysiology, and behavior are merely aspects of the same reality viewed from different disciplines, identifying biological substrates of behavior is no different from identifying behavioral markers of biology. Neither does much to resolve the essential philosophical and legal problems inherent in the free will-determinism dilemma. As moral persons, we may still be forced to judge others personally as though there were absolute determinism and ourselves as though there were absolute free will while legally attempting to reconcile punishment with protection.

---

*Note Added in Press to Page 43. Since this writing, the 5-HT3 receptor has not only been identified as a ligand-gated ion channel (Yalkel, J. L. and M. B. Jackson, 1988 *Neurone* 1:615; Derkach, V., A. Surprenant, and R. North, 1989 *Nature* 339: 706), and a host of selective ligands developed (Richardson, B. P. and G. Engle, 1986 *Trends in Neurosci.*, 9: 424), but it has been cloned, its primary structure elucidated, and the regional distribution of its messenger RNA defined (Maricq, A. V., A. S. Peterson, A. J. Brake, R. M. Meyers, and D. Julius, 1991 *Science*, 254: 432).

5

Serotonin and Suicide

Dan J. Stein
and
Michael Stanley

Abstract: There is evidence for decreased serotonin function in attempted and successful suicide, but this correlation needs to be related to psychosocial causal mechanisms.

This chapter addresses the neurobiology of suicide, highlighting in particular research on the serotonin neurotransmitter system. Studies of the relationship between suicide and serotonin are to some extent presented in historical sequence, beginning with the earliest findings and moving to the most recent. Thus the chapter moves from the earlier research focus on depression to the later focus on impulsivity. As the focus shifts, so do the issues raised in this chapter move from the more biological to the more sociological and legal.

MODERN PSYCHIATRY AND NEUROTRANSMISSION

Biological suicidology may be understood as originating in modern psychiatry's investigation of the arena of overlap between psychiatric disorder and neurotransmission. In the 1950s, the first hypotheses of a relationship between psychiatric disorder and neurotransmitter systems were made. In view of the chemical similarity between serotonin and some hallucinogenic drugs (e.g., LSD) and of the clinical similarity between schizophrenia and hallucinogen intoxication, it was suggested that serotonin was etiologically relevant to schizophrenia (Woolley and Shaw, 1954).

At the same time, the serendipitous discovery that certain drugs, such as chlorpromazine, were clinically useful in psychiatric disorders (e.g., psychosis) gave researchers the opportunity to investigate the mechanism of action of effective psychotropics. The finding that the effect of psychotropics is exerted by changes in specific chemical processes in neuronal transmission cemented this relationship between psychiatric disorders and neurotransmitter systems.

Neurotransmitters may be defined as chemical messengers, which allow the transfer of information between neurons. These messengers are released by the presynaptic neuron, diffuse across the synaptic cleft, and bind with receptor sites on the postsynaptic membrane. Receptor sites serve as a recognition mechanism for various neurotransmitters. That is, there are receptor sites in the brain that selectively and either preferentially or exclusively recognize one neurotransmitter over others. Other events that play a role in the system include synthesis of the messenger by the presynaptic neuron, its degradation by enzymes in the synaptic cleft, and its uptake by presynaptic receptors for reuse. Presynaptic nerve cells may release neurotransmitters into the synaptic cleft to initiate either an excitatory or an inhibitory response in the targeted postsynaptic neuron. The neurotransmitters are of different types of molecules, including amino acids, neuropeptides, and the biogenic amines. Neurons that use each of the monoamines dopamine, serotonin, and noradrenaline are thought to comprise separate anatomic pathways that possibly serve distinct functions. It was therefore possible to hypothesize that specific changes in neurotransmitter systems were involved in the etiology and treatment of particular psychiatric disorders.

In figure 4.1, a serotonergic synapse is depicted indicating the specific serotonin postsynaptic receptor sites. Also sketched are the synthesis of serotonin from tryptophan, the degradation of serotonin by the monoamine oxidase enzyme (MAO), and the reuptake of serotonin by presynaptic receptors (see the previous chapter).

Research methodologies for examining the monoamine neurotransmitter systems in psychiatric patients have been varied. Such processes as synthesis and metabolic turnover of neurotransmitters have been investigated by measuring levels of neurotransmitters and their precursors, metabolites, and metabolic enzymes in body fluids, such as plasma, urine, and cerebrospinal fluid (CSF), or in postmortem brain tissue. The former measurements may be confounded by the contribution of peripheral sources of the same neurotransmitters and do not allow the assessment of localized abnormalities. The latter measurements may be confounded by changes at death. Such processes as receptor function have been investigated by employing radioactive ligands or drugs that bind with specificity and affinity to receptor systems to measure the number of receptor sites and their binding affinity, or by administering specific receptor agonists or antagonists that have characteristic effects on receptor systems to test whether the system is hyporesponsive or hyperresponsive. In interpreting any of the results of these investigations one needs to bear in mind that neurotransmitter systems are dynamic and are affected by such factors as circadian rhythms, diet, medication, and clinical status.

Three decades of research into neurotransmitter processes and psychiatric disorders have provided psychiatry with theoretically important hypotheses regarding the etiology of psychiatric illness, as well as clinically useful hypotheses concerning its psychotropic treatment. This research comprises the cornerstone of biological psychiatry. In the next section, we move to a particular focus of such research, namely, depression.

SEROTININ, SUICIDE, AND DEPRESSION

In the 1960s, it was discovered that administering reserpine, which depletes brain stores of the monoamine neurotransmitters serotonin and noradrenaline, results in depression. Conversely, it was found that effective antidepressants decrease monoamine turnover—the tricyclic drugs inhibit monoamine uptake, and the MAO inhibitors decrease enzymatic degradation of monoamines. The pharmacologic effect of both classes of drugs was a net increase in the availability of neurotransmitters within the synaptic cleft. In consequence, the monoamine hypothesis of depression was formulated. Researchers held that diminished neurotransmission of monoamines was etiologically relevant to depression—workers in the United States favored diminished noradrenaline transmission, while those in Europe favored diminished serotonin transmission (Schildkraut, 1965; Lapin and Oxenkrug, 1969).

Although the monoamine theory of depression has subsequently turned out to be oversimplistic, it generated a vast amount of important research. To prove the hypothesis, it was necessary to examine monoamine neurotransmitters (and their precursors, metabolites, enzymes, and receptors) in postmortem subjects and with in vivo studies. As reviewed in the following sections, a correlation emerged from these studies between serotonin measurements and suicidal behavior.

Postmortem Studies

The first postmortem studies of neurotransmitters in suicide victims were carried out in the late 1960s. Measurements were made of serotonin, noradrenaline, and dopamine and of their respective metabolites, 5-hydroxyindoleacetic acid (5-HIAA), 3-methoxy-4-hydroxyphenylglycol (MHPG), and homovanillic acid (HVA). Early findings were that brain-stem levels of serotonin, or 5-HIAA, are decreased in suicide victims (Shaw et al, 1967; Bourne et al, 1968; Pare et al, 1969). Lloyd et al (1974) went on to measure serotonin and 5-HIAA in even more localized fashion, in discrete nuclei in the raphe region of the brain stem, the location of most of the brain's serotonergic cell bodies. They found significantly lower serotonin concentrations in the dorsal raphe nucleus and in the nucleus centralis inferior. The majority of subsequent studies (Beskow et al, 1976; Cochran et al, 1976; Korpi et al, 1983; Stanley et al, 1983; Owens et al, 1983; Crow et al, 1984; Kleinman, 1985) have confirmed the general finding of decreased serotonin and/or 5-HIAA in brain-stem (raphe nuclei) and/or subcortical nuclei (hypothalamus) as the most consistent postmortem changes in these measurements in this population. That is, postmortem studies show that individuals who commit suicide have lower levels of serotonin or its byproduct (5-HIAA) in critical regions of the brain than do normal individuals. Since no abnormalities have been consistently found for other monoamines (noradrenaline, dopamine), serotonin level seems to be associated in some way with suicide. Such evidence, however, does not establish that serotonin malfunction is causally related to suicide.

A number of methodological problems may be noted in these studies. Such intervening variables as the cause of death, postmortem interval, and acute drug or alcohol use may influence results. Researchers therefore turned to receptor-binding studies; binding is thought to be less responsive to acute changes (Peroutka and Snyder, 1980). Results of such studies in general provided evidence for the validity of the earlier postmortem findings. Stanley et al (1982) found reduced imipramine binding in the brains of suicide victims, a finding replicated in the majority of subsequent studies (Meyerson et al, 1982; Perry et al, 1983; Paul et al, 1984; Crow et al, 1984; Gross-Isseroff et al, 1989). Imipramine binding is thought to be associated with presynaptic neurons, and reduced imipramine binding may indicate a decrease in functional serotonergic neurons, with consequent reduction in serotonin release. Further, Stanley et al (1983) found an increase in 5-hydroxytryptamine, subtype 2 (5-HT$_2$), binding in suicide victims, a finding again replicated in the majority of subsequent studies (Owens et al, 1983; Mann et al, 1986; Meltzer et al, 1987; Cheetam et al, 1988; Arora and Meltzer, 1989). Binding of 5-HT$_2$ is thought to be associated with postsynaptic receptors, and increased 5-HT$_2$ binding may indicate the brain's reaction to a decrease in functional serotonergic neurons, with consequent augmentation (up-regulation) of serotonin recognition sites. This hypothesis was further strengthened by the finding that imipramine binding and 5-HT$_2$ binding are negatively correlated, as would be expected if reduced serotonin release led to some compensatory increase in postsynaptic serotonergic receptor sites.

The findings from the receptor-binding studies are consistent with the results of those studies that reported reductions in serotonin and its metabolite 5-HIAA. That is, a decrease in presynaptic binding sites (as measured by imipramine) would coincide with a reduction in serotonin levels. If there are fewer sites releasing serotonin, there will be less serotonin to measure. As a general consequence of reduced serotonin being released presynaptically, postsynaptic receptor sites might proliferate in number, compensating for the diminished supplies of serotonin in order to augment its signal.

Thus, reduced levels of serotonin may lead to decreases in presynaptic release sites, which could lead to an increase in postsynaptic recognition sites. The preceding changes may in turn result in the biochemical conditions that would favor suicidal behaviors.

In Vivo Studies

In the early 1970s, the first clinical studies of neurotransmitter metabolites in the CSF of depressed patients were undertaken. Despite the many influences on CSF 5-HIAA and the necessity for sensitive and specific analytic methods (Asberg et al, 1986), such studies may be performed with more control of variables than is possible in postmortem studies. In addition, Stanley et al (1985) have provided strong evidence that CSF 5-HIAA does in fact reflect central serotonin activity.

Asberg et al (1976a) did not find a correlation between low CSF 5-HIAA and depression, and subsequent studies have confirmed that the relationship between

CSF 5-HIAA and depression is subtle at best. Post et al (1980) have reviewed 15 studies and have found that on average the CSF 5-HIAA in depressed subjects is 78% that of nondepressed controls. Only 30% of depressed patients have significantly lower concentrations of CSF 5-HIAA. However, in attempting to explain their results, Asberg et al (1976b) compared low 5-HIAA with other variables and showed that in subjects with low 5-HIAA there was an increased incidence of committed or attempted suicide by violent means. Subsequent studies on populations from Scandinavia (Agren, 1980), the United States (Leckman et al, 1981), Holland (van Praag, 1982), Britain (Montgomery and Montgomery, 1982), India (Palanappian et al, 1983), Hungary (Banki et al, 1984), and Spain (Lopez-Ibor et al, 1985) have in general confirmed that suicidal, depressed patients have lower CSF 5-HIAA than do nonsuicidal, depressed patients. Further, those studies that do not confirm this finding often have a predominance of patients with bipolar disorder (Roy-Byrne et al, 1983; Berrettini et al, 1986), in which serotonergic dysfunction may be primary to the illness itself and therefore may obscure the relationship between suicide and serotonin (Goodwin, 1986).

As mentioned earlier, the precise relationship between peripheral serotonin measures (e.g., platelet 5-HIAA uptake and content and platelet MAO activity) and central serotonergic activity is presently unclear. While CSF levels of 5-HIAA have been shown to correlate with 5-HIAA in the brain (Stanley et al, 1985; Knott et al, 1989), no such relationship has been reported between central and peripheral measures of serotonergic activity. For example, imipramine binding in the platelet, a peripheral marker for serotonergic function, is not correlated with CSF levels of 5-HIAA, a central serotonergic marker. Furthermore, studies of the same measures in depressed and suicidal patients have yielded contradictory results. Nevertheless, in the case of platelet MAO activity, a number of studies seem to indicate at least some evidence of the absence of an association with depression, but they indicate the presence of a negative correlation with suicide ratings that is possibly consistent with diminished presynaptic serotonergic activity (Buchsbaum et al, 1979; Gottfries et al, 1980; Meltzer and Arora, 1986).

Work done using pharmacological challenges that activate the serotonergic system has been consistent with the results reviewed above. Coccaro et al (1989) report that in patients with depression a reduced prolactin response to fenfluramine correlates with a history of attempted suicide. Fenfluramine is an indirect serotonin agonist that releases endogenous stores of serotonin from presynaptic neurons and inhibits its reuptake. Therefore, prolactin release is the result of net serotonin release, and this finding is consistent with the hypothesis of a net decrease in serotonin transmission in suicidal patients. Meltzer et al (1984) reported an enhanced cortisol response to 5-hydroxytryptophan (5-HTP), a serotonin precursor, in depressed patients who had made a suicide attempt compared to patients who had not. Meltzer et al (1987) obtained a similar result using the serotonin agonist MK212, which acts directly on serotonin receptors. Such findings are in line with the hypothesis of postsynaptic serotonin receptor hyperresponsivity in suicide victims.

It should be emphasized that suicide represents a diagnostically heterogeneous group of patients. The initial postmortem studies of suicide victims made the assumption that suicide was associated with depression, and to some extent this is accurate, for there is much evidence that a high proportion of suicides have a primary depressive illness (Sainsbury, 1986). Nonetheless, it is clear that suicide and depression are neither synonymous nor necessarily coexistent states. There has thus been a call for the study of postmortem brains in depressed patients who have died through methods other than suicide (Ferrier et al, 1986). Inversely, there is a need to study suicidal patients who are not depressed. This concept provides a link to the next section of this chapter, which presents those findings that confirm the correlation between serotonin measurements and suicidal behavior, rather than depression per se.

SEROTONIN, SUICIDE, AND IMPULSIVITY

If the monoamine hypothesis of depression was a driving principle behind research into serotonin and psychopathology in the 1970s, it may be that in the 1990s the driving hypothesis will concern the interrelations between serotonin and impulsivity and compulsivity. Impulsive and compulsive personality styles can be viewed as behavioral opposites. While compulsive individuals can move with excessive deliberateness, the impulsive individual can move from one behavior to the next with rapidity and little forethought. It therefore becomes of particular interest that both of these behavioral extremes appear to be mediated by serotonergic mechanisms.

Such a hypothesis has gradually come to be formulated over the past decade. The link between serotonin and impulsivity rests on a number of studies. Brown et al (1979, 1982) demonstrated a decrease in CSF 5-HIAA in patients with severe personality disorder, a condition in which impulsivity is a characteristic feature. The decrease in CSF 5-HIAA and scores on a lifetime aggression scale were correlated. Linnoila et al (1983, and chapter 6, this volume) showed a decrease in CSF 5-HIAA in murderers and attempted murderers, a decrease that correlated with violence that was assessed as impulsive. Schalling et al (1984) have reported an inverse correlation between CSF 5-HIAA and scores on the impulsiveness scale of the Karolinska Scales of Personality in patients and healthy volunteers. Later studies in general confirm this trend (Lidberg et al, 1985; Virkkunen et al, 1987; Roy et al, 1988). Other measures of serotonergic activity are consistent with the CSF findings; thus, platelet MAO activity (Schalling et al, 1975, 1988; von Knorring et al, 1984; Perris et al, 1984) and platelet serotonin uptake (Brown et al, 1989) are negatively correlated with impulsivity in a variety of subject populations. Less rigorous evidence lies in reports of the efficacy of carbamazepine, lithium, and propranolol—all of which have effects on the serotonin system—in the treatment of impulsive aggression. More rigorous evidence lies in the use of pharmacological challenges with fenfluramine (Coccaro et al, 1989) to demonstrate reduced central serotonergic function correlating with a history of impulsive aggression in patients with personality disorder. (A variety of animal studies [Soubrie, 1986], which we will not cover here, have been adduced to underpin the serotonin hypothesis of impulsivity.)

The link between serotonin and compulsivity is based on the finding that obsessive-compulsive disorder responds to medication that acts selectively on the serotonin system. Patients who respond to clorimipramine, a tricyclic antidepressant with relatively specific serotonin reuptake-blocking effects, have higher baseline levels of CSF 5-HIAA and platelet serotonin than do nonresponders, and improvement in their obsessive-compulsive symptoms correlates with normalization of these measures during treatment (Thoren et al, 1980; Flament et al, 1985). Finally, m-chlorophenylpiperazine (m-CPP), a selective serotonin receptor agonist, exacerbates obsessive-compulsive symptoms prior to treatment with clomipramine but not after (Zohar et al, 1988; Hollander et al, 1990), a finding consistent with the hypothesis that in the treatment of obsessive-compulsive disorder a hyperresponsive serotonin system is down-regulated by chronic administration of a serotonergic medication.

The role of serotonin in impulsivity may account for the relationship between serotonin and suicide. Indeed, studies of men with personality disorder (Brown et al, 1979, 1982) and a study of violent offenders (Linnoila et al, 1983) also demonstrated lower CSF 5-HIAA in subjects with suicidal behavior. Having committed a murder is perhaps the strongest risk factor for suicide (West, 1965); and Lidberg et al (1984) found very low CSF 5-HIAA in three cases of suicide attempters who had killed their own children. Conversely, in studies by Traskman et al (1981), van Praag (1982, 1983), and Banki et al (1984) on suicidal patients with a variety of diagnoses, it is apparent that the association between CSF 5-HIAA and suicidal behavior holds particularly in those with violent suicide attempts. In their work using fenfluramine challenges to establish reduced serotonergic function, Coccaro et al (1989) show that reduced neuroendocrine response correlates with a history of suicide attempt in both depressed and personality disorder patients but with impulsive aggression in personality disorder patients only.

It is generally accepted that such a concept as depression comprises a variety of disorders with disparate subtypes. Similarly, impulsivity and aggression are concepts that require further dissection. Distinctions need to be drawn (for example, impulsivity is different from violence, which may be premeditated) and intersections need to be elaborated (for example, both psychoanalysts [Freud, 1957] and neurobiologists [van Praag et al, 1986] have argued that aggression and depression are intimately linked).

Nevertheless, studies of impulsivity do seem to indicate that the relationship between suicide and serotonin is independent of diagnostic borders, such as depression. Further work that supports this conclusion is that which shows lower CSF 5-HIAA in suicide victims who had such diverse diagnoses as schizophrenia (van Praag, 1983; Ninan et al, 1984; Banki et al, 1984) and alcoholism (Banki et al, 1984). Although the work on schizophrenia has not yet been uniformly replicated (Roy et al, 1985; Pickar et al, 1986; Stanley, 1986) and serotonergic dysfunction in alcoholism (Ballenger et al, 1979) has not yet been fully clarified, such data reinforce the significance of the neurobiological correlation of suicide and serotonin as a finding that is independent of previously observed psychosocial risk factors for suicide.

There are of course many aspects of the relationship between suicide and serotonin that remain unclear. Just as modern psychiatry has failed to find a model of neurotransmission in a psychiatric disorder that is as clear-cut as the neurological model of dopamine deficiency in Parkinson's disease, so biological suicidology remains mired in complexity. Thus, for example, it is clear that serotonin is a neurotransmitter system with multiple receptors; but the particular role of each of these receptors is presently unknown, and many of the studies reviewed above need to be replicated with the more specific ligands that have become available. Moreover, the ways in which serotonin interacts with other neurotransmitter and neuroendocrine systems require elucidation. Conversely, serotonin is affected in a variety of disorders other than simply in depression and impulsive and compulsive disorders (van Kammen, 1987); thus the specificity of the serotonin disturbance in suicide remains unclear.

SOCIOLOGICAL AND LEGAL IMPLICATIONS

The basis for this volume is the possible influence of neurobiological knowledge (in particular, of neurotransmitters and their influence on behavior) on society and law. In line with a general trend in American psychiatric research, it is clear that the neurobiological study of suicide has made major advances in the past few decades, adding considerably to an older literature on the psychosocial aspects of suicide. Researchers in this arena have begun to question the ethical and other implications of their work. Stanley (1986) asks what patients with biological markers for suicide should be told and wonders whether these patients will be entitled to obtain medical insurance. New neurobiological knowledge has the potential to lead to a revision of our concept of what persons are and what behaviors signify and entail and of our moral and ethical responses to people and their actions. The work on serotonin and suicide exemplifies these issues and can perhaps help illuminate them.

In this section we point out a number of limits to the findings of neurobiological suicidology. The first limit concerns the matter of prediction. It has been shown that among patients who have made a suicide attempt, those with low CSF serotonin have a 10-times' greater chance of succeeding (Asberg et al, 1986). Nevertheless, even in those psychiatric patients at greatest risk for suicide (past attempters, depressives, schizophrenics), suicide is a rare event, with a prevalence rate of only 1% to 2% per year (Pokorny, 1983), making prediction difficult. As Cohen (1986) points out, for a putative superpredictor with 90% sensitivity and 90% specificity, the false-positive rate is 92%. If a population had a rate of suicide of 1% and this predictor were used to screen 10,000 patients, then 1080 predicted suicides would be identified, of whom 990 would be false positives (i.e., individuals identified as being at risk who in fact do not commit suicide). The validity of the predictor, expressed as a percentage of correctly predicted outcomes, corrected for chance (kappa), would then be 0.137. Thus the correlation of low CSF serotonin and suicide, although statistically significant, is of little practical significance by itself.

One may argue that, although neurobiological studies of suicide do not enable suicide prediction, they provide an understanding of the neurobiological underpinnings of suicidal behavior. This in turn may suggest the use of particular psychotropic drugs in the treatment of suicidal behavior. However, the studies that have been reviewed in this chapter indicate a correlation between measurements of serotonin and suicidal behavior, and this should not be confused with causal explanation. The paucity of causal explanations in the neurobiology of suicide would seem to constitute a second limit.

It may be objected that while the findings presented here are admittedly incomplete, they do in fact provide the beginnings of a detailed neurobiology of suicidal behavior. Nevertheless, we wish to establish a third limit, and in order to do so we can do no better than to return to the work of Émile Durkheim. Durkheim (1951) reminds us that suicide denotes an intent to die. "We may then say conclusively: The term suicide is applied to all cases of death resulting directly or indirectly from a positive or negative act of the victim himself which he knows will produce this result. An attempt is an act thus defined but falling short of actual death." Although the subject of intent is usually tackled by those who possess greater philosophical sophistication, we feel that, at least insofar as we restrict our comments to the subject of suicide, we are entitled to make some observations.

It is clear from this review that certain causal mechanisms of suicide have in fact begun to be described. They center on the clinical syndrome of depression and the psychometric pattern of impulsivity, as well as on their neurobiological underpinnings. Now, our knowledge of this neurobiology will undoubtedly increase. The explanation of the intention to kill oneself nevertheless entails more than the elaboration of these etiologically relevant behavior patterns and their neurobiological substrates. Thus, there are those people who kill themselves without appearing depressed or impulsive, and there are those who are depressed and impulsive who do not kill themselves. The complexity of an intention, such as aiming to kill oneself, is invariably such that no single causal mechanism will explain it. The causal mechanisms employed to explain the intent to commit suicide must include both neurobiological and psychosocial ones. Furthermore, such mechanisms need not necessarily be restricted to the realm of pathology—more and more it is being argued that in certain circumstances the processes that lead a person to commit suicide may conceivably be rational (Szasz, 1971; Motto, 1981; Mayo, 1983).

To explain the intent to commit suicide, then, the causal mechanisms that we employ must include both neurobiological and psychosocial ones, pathological and normal ones. Different kinds of causal mechanisms of suicide may allow the differentiation of different populations (von Andics, 1938; Linehan, 1986) of patients with suicidal ideation. Consequent distinctions in etiological theory and therapeutic response are required. To the extent that work on serotonin and suicide provides only a partial description of the causal mechanisms of the intent to kill oneself, this work is intrinsically limited.

We have reviewed the findings of neurobiological suicidology and have concluded that they may help elucidate the causal mechanisms of suicidal behavior. We have earlier remarked that many aspects of such work require further investigation, but we are optimistic also that further research will advance the field. However, in this section we have pointed to the limits of this work, limits that establish the necessity for incorporating a thorough psychosocial knowledge base in the consideration of both the etiology and the treatment of suicide. Future research needs to combine the investigation of neurobiological measurements, psychosocial factors, and the intent of patients. As Shapiro (chapter 12, this volume) points out, it is certainly useful to consider how new developments in neurobiological knowledge may change our view of people and morals. But in doing so, we should be clear in establishing the present empirical limits of our science, as well as the future theoretical ones. Jeffery's claim (chapter 11, this volume) that a person's moral self is defined by his or her serotonin level seems to us in general, and certainly as regards suicidal behavior, both empirically untrue at present and probably theoretically invalid for the future.

REFERENCES

Agren H: Symptom patterns in unipolar and bipolar depression correlating with monoamine metabolites in the cerebrospinal fluid. II. Suicide. *Psychiatry Res* 3:226–236, 1980.

Arora RC, Meltzer HY: Serotonergic measures in the brains of suicide victims: 5-HT$_2$ binding sites in the frontal cortex of suicide victims and control subjects. *Am J Psychiatry* 146:730–736, 1989.

Asberg M, Nordstrom P, Traskman-Bendz L: Biological factors in suicide, in *Suicide*. Edited by Roy A. Baltimore, Williams & Wilkins, 1986.

Asberg M, Thoren P, Traskman L, et al: Serotonin depression: a biochemical subgroup within the affective disorders? *Science* 191:478–480, 1976a.

Asberg M, Traskman L, Thoren P: 5-HIAA in the cerebrospinal fluid—a biochemical suicide predictor? *Arch Gen Psychiatry* 33:1193–1197, 1976b.

Ballenger JC, Goodwin FK, Major LF, et al: Alcohol and central serotonin metabolism in man. *Arch Gen Psychiatry* 36:224–227, 1979.

Banki CM, Arato M, Papp Z, et al: Biochemical markers in suicidal patients: investigations with cerebrospinal fluid amine metabolites and neuroendocrine tests. *J Affective Disord* 6:341–350, 1984.

Berrettini W, Nurenberger J, Narrow W, et al: Cerebrospinal fluid studies of bipolar patients with and without a history of suicide attempts, in *Psychobiology of Suicidal Behavior*. Edited by Mann JJ, Stanley M. New York, New York Academy of Sciences, 1986.

Beskow J, Gottfries CG, Roos BE, et al: Determination of monoamine and monoamine metabolites in the human brain; postmortem studies in a group of suicides and in a control group. *Acta Psychiatr Scand* 53:7–20, 1976.

Bourne HR, Bunney WE Jr, Colburn RW, et al: Noradrenaline, 5-hydroxytryptamine, and 5-hydroxyindoleacetic acid in the hindbrains of suicidal patients. *Lancet* 2:805–808, 1968.

Brown CS, Kent TA, Bryant SG, et al: Blood platelet uptake of serotonin in episodic aggression. *Psychiatry Res* 27:5–12, 1989.

Brown G, Ebert M, Goyer P, et al: Aggression, suicide, and serotonin. *Am J Psychiatry* 139:741–746, 1982.

Brown G, Goodwin F, Ballenger J, et al: Aggression in humans correlates with cerebrospinal fluid amine metabolites. *Psychiatry Res* 1:131–139, 1979.

Buchsbaum MS, Hairr RJ, Murphy DL: Suicide attempts, platelet monoamine oxidase and the average evoked response. *Acta Psychiatr Scand* 56:69–77, 1979.

Cheetam SC, Crompton MR, Katona CL, et al: Brain 5-HT$_2$ receptor binding sites in depressed suicide victims. *Brain Res* 443:272–280, 1988.

Coccaro EF, Siever LJ, Klar HM, et al: Serotonergic studies in patients with affective and personality disorders: correlates with suicidal and impulsive aggressive behavior. *Arch Gen Psychiatry* 46:587–599, 1989.

Cochran E, Robins E, Grate S: Regional serotonin levels in the brain: a comparison of depressive suicides and alcoholic suicides with controls. *Biol Psychiatry* 11:283–294, 1976.

Cohen J: Statistical approaches to suicidal risk factor analysis, in *Psychobiology of Suicidal Behavior*. Edited by Mann JJ, Stanley M. New York, New York Academy of Sciences, 1986.

Crow TJ, Cross AJ, Cooper SJ, et al: Neurotransmitter receptors and monoamine metabolites in the patients with Alzheimer-type dementia and depression and suicides. *Neuropharmacology* 23:1561–1569, 1984.

Durkheim É: *Suicide*. Glencoe, The Free Press, 1951. (Originally published as *Le Suicide*, 1897.)

Ferrier IN, McKeith IG, Cross AJ, et al: Postmortem neurochemical studies in depression, in *Psychobiology of Suicidal Behavior*. Edited by Mann JJ, Stanley M. New York, New York Academy of Sciences, 1986.

Flament MF, Rapoport JL, Berg CL, et al: Clomipramine treatment of childhood obsessive-compulsive disorder: a double-blind controlled study. *Arch Gen Psychiatry* 42:977–986, 1985.

Freud S: Mourning and melancholia, in *The Standard Edition of the Complete Psychological Works of Sigmund Freud*. Edited by Strachey J. London, Hogarth Press, 1957.

Goodwin FK: Suicide, aggression and depression: a theoretical framework for future research, in *Psychobiology of Suicidal Behavior*. Edited by Mann JJ, Stanley M. New York, New York Academy of Sciences, 1986.

Gottfries CG, Knorring LV, Oreland L: Platelet monoamine oxidase activity in mental disorders. *Neuropsychopharmacology* 4:185–192, 1980.

Gross-Isseroff R, Israeli M, Biegon A: Autoradiographic analysis of tritiated imipramine binding in the human brain post mortem: effects of suicide. *Arch Gen Psychiatry* 46:237–241, 1989.

Hollander E, DeLaria C, Gully R, et al: Effects of chronic fluoxetine treatment on behavioral and neuroendocrine responses to meta-chloro-phenylpiperazine in obsessive-compulsive disorder. *Psychiatry Res* 36:1–17, 1990.

Kleinman J: Muscarinic receptor density in skin fibroblast and autopsied brain tissue in affective disorder. Paper presented at Psychobiology of Suicide Behavior. New York, New York Academy of Sciences, September 1985.

Knott P, Haroutuniam V, Bierer L, et al: Correlations post-mortem between ventricular CSF and cortical tissue concentrations of MHPG, 5-HIAA and HVA in Alzheimer's disease. *Biol-Psychiatry* 111A–119A, 1989.

Korpi ER, Kleinman JE, Goodman SI, et al: Serotonin and 5-hydroxyindoleacetic acid concentrations in different brain regions of suicide victims: comparison in chronic schizophrenic patients with suicide as cause of death. Presented at the International Society for Neurochemistry, Vancouver, Canada, 1983.

Lapin IP, Oxenkrug GF: Intensification of the central serotonergic processes as a possible determinant of the thymoleptic effect. *Lancet* 1:132–136, 1969.

Leckman JF, Charney DS, Nelson CR, et al: CSF tryptophan, 5-HIAA and HVA in 132 patients characterized by diagnosis and clinical state. *Recent Adv Neuropsychopharmacology* 31:289–297, 1981.

Lidberg L, Asberg M, Sundquist-Stensman UB: 5-hydroxyindoleacetic acid levels in attempted suicides who have killed their children. *Lancet* 2:928, 1984.

Lidberg L, Tuck JR, Asberg M, et al: Homicide, suicide and CSF 5-HIAA. *Acta Psychiatr Scand* 71:230–236, 1985.

Linehan MM: Suicidal people: one population or two? in *Psychobiology of Suicidal Behavior*. Edited by Mann JJ, Stanley M. New York, New York Academy of Sciences, 1986.

Linnoila M, Virkkunen M, Scheinin M, et al: Low cerebrospinal fluid 5-hydroxyindoleacetic acid concentrations differentiates impulsive from non-impulsive violent behavior. *Life Sciences* 33:2609–2614, 1983.

Lloyd KG, Farley IJ, Deck JHN, et al: Low cerebrospinal fluid 5-hydroxyindoleacetic acid in discrete areas of the brainstem of suicide victims and control patients, in *Advances in Biochemical Psychopharmacology* 2:387–397. New York, Raven Press, 1974.

Lopez-Ibor JJ, Saiz-Ruiz J, Perez de los Cobos JC: Biological correlations of suicide and aggressivity in major depressions (with melancholia): 5-hydroxyindoleacetic acid and cortisol in cerebral spinal fluid, dexamethasone suppression test and therapeutic response to 5-hydroxytryptophan. *Neuropsychobiology* 14:67–74, 1985.

Mann JJ, Stanley M, McBride AP, et al: Increased serotonin and beta-adrenergic receptor binding in the frontal cortices of suicide victims. *Arch Gen Psychiatry* 43:954–959, 1986.

Mayo D: Contemporary philosophical literature on suicide. *Suicide Life Threatening Behavior* 13:313–345, 1983.

Meltzer HY, Arora RC: Platelet markers of suicidality, in *Psychobiology of Suicidal Behavior*. Edited by Mann JJ, Stanley M. New York, New York Academy of Sciences, 1986.

Meltzer HY, Nash JF, Ohmori T, et al: Neuroendocrine and biochemical studies of serotonin and dopamine in depression and suicide. International Conference on New Directions in Affective Disorders: *Book of Abstracts*, 1987.

Meltzer HY, Perline R, Tricou BJ, et al: Effect of 5-hydryoxytryptophan on serum cortisol levels in major affective disorders, II: relation to suicide, psychosis and depressive symptoms. *Arch Gen Psychiatry* 41:379–387, 1984.

Meyerson LR, Wennogle LP, Abel MS, et al: Human brain receptor alterations in suicide victims. *Pharmacol Biochem Behav* 17:159–163, 1982.

Montgomery SA, Montgomery D: Pharmacological prevention of suicidal behavior. *J Affective Disord* 4:291–298, 1982.

Motto J: Rational suicide and medical ethics, in *Rights and Responsibilities in Modern Medicine*. Edited by Basson M. New York, Alan R Liss Inc, 1981.

Ninan PT, van Kammen DP, Scheinin M, et al: CSF 5-hydroxyindoleacetic acid in suicidal schizophrenic patients. *Ma J Psychiatry* 141:566–569, 1984.

Owens F, Cross AJ, Crow TJ, et al: Brain 5-HT$_2$ receptors and suicide. *Lancet* 2:1256, 1983.

Palanappian V, Ramachandran V, Somasundaram O: Suicidal ideation and biogenic amines in depression. *Indian J Psychiatry* 25:286–292, 1983.

Pare CMB, Yeung DPH, Price K, et al: 5-hydroxytryptamine, noradrenaline, and dopamine in brainstem, hypothalamus, and caudate nucleus of controls and of patients committing suicide by coal-gas poisoning. *Lancet* 1:131–135, 1969.

Paul SM, Rehavi M, Skolnick P, et al: High affinity binding of antidepressants to a biogenic amine transport site in human brain and platelet: studies in depression, in *Neurobiology and Mood Disorders*. Edited by Post RM, Ballenger JC. Baltimore, Williams & Wilkins, 1984.

Peroutka SJ, Snyder SH: Regulation of serotonin (5-HT$_2$) receptors labeled with [^{3}H] spiroperidol by chronic treatment with antidepressant amitriptyline. *Pharmacol Exp Ther* 215:582–587, 1980.

Perris C, Eisemann M, von Knorring L, et al: Personality traits and monoamine oxidase activity in platelets in depressed patients. *Neuropsychobiology* 12:201–205, 1984.

Perry EK, Marshall EF, Blessed G, et al: Decreased imipramine binding in the brains of patients with depressive illness. *Br J Psychiatry* 142:188–192, 1983.

Pickar D, Roy A, Breier A, et al: Suicide and aggression in schizophrenia: neurobiologic correlates, in *Psychobiology of Suicidal Behavior*. Edited by Mann JJ, Stanley M. New York, New York Academy of Sciences, 1986.

Pokorny AD: Prediction of suicide in psychiatric patients. *Arch Gen Psychiatry* 40:249–257, 1983.

Post RM, Ballenger JC, Goodwin FK: Cerebrospinal fluid studies of neurotransmitter function in manic and depressive illness, in *Neurobiology of Cerebrospinal Fluid*. Edited by Wood JH. New York, Plenum Press, 1980.

Roy A, Adinoff B, Linnoila M: Acting out hostility in normal volunteers: negative correlation with levels of 5-HIAA in cerebrospinal fluid. *Psychiatry Res* 24:187–194, 1988.

Roy A, Ninan P, Mazonson A, et al: CSF monoamine metabolites in chronic schizophrenic patients who attempt suicide. *Psychol Med* 15:335–340, 1985.

Roy-Byrne P, Post RM, Rubinow DR, et al: CSF 5-HIAA and personal and family history of suicide in affectively ill patients: a negative study. *Psychiatry Res* 10:263–274, 1983.

Sainsbury P: Depression, suicide, and suicide prevention, in *Suicide*. Edited by Roy A. Baltimore, Williams & Wilkins, 1986.

Schalling D, Asberg M, Edman G, et al: Impulsivity, nonconformity and sensation seeking as related to biological markers for vulnerability. *Clin Neuropharmacol* 7:746–747, 1984.

Schalling D, Cronholm B, Asberg M: Components of state and trait anxiety as related to personality and arousal, in *Emotions, Their Parameters and Measurement*. Edited by Levi L. New York, Raven Press, 1975.

Schalling D, Edman G, Asberg M, et al: Platelet MAO activity associated with impulsivity and aggressivity. *Person Individ Diff* 9:597–605, 1988.

Schildkraut JJ: The catecholamine hypothesis of affective disorders: a review of supporting evidence. *Am J Psychiatry* 122:509–522, 1965.

Shaw DM, Camps FE, Eccleston EG: 5-hydroxytryptamine in the hind-brain of depressive suicides. *Br J Psychiatry* 113:1407–1411, 1967.

Soubrie P: Reconciling the role of central serotonin neurones in human and animal behavior. *Behav Brain Sci* 9:319–364, 1986.

Stanley B: Ethical considerations in biological research on suicide, in *Psychobiology of Suicidal Behavior*. Edited by Mann JJ, Stanley M. New York, New York Academy of Sciences, 1986.

Stanley M, McIntyre I, Gershon S: Postmortem serotonin metabolism in suicide victims. Paper presented at the American College of Neuropsychopharmacology, Puerto Rico, 1983.

Stanley M, Mann JJ: Increased serotonin-2 binding sites in frontal cortex of suicide victims. *Lancet* 2:214–216, 1983.

Stanley M, Stanley B, Traskman-Bendz L, et al: Depressive symptoms and CSF levels of 5-HIAA in schizophrenic patients who have attempted suicide. *American College of Neuropsychopharmacology*, 1986.

Stanley M, Traskman-Bendz L, Dorovini-Zis K: Correlations between aminergic metabolites simultaneously obtained from human CSF and brain. *Life Sciences* 37:1279–1286, 1985.

Stanley M, Virgilio J, Gershon S: Triated imipramine binding sites are decreased in the frontal cortex of suicides. *Science* 216:1337–1339, 1982.

Szasz T: The ethics of suicide. *Antioch Rev* 31:7–17, 1971.

Thoren P, Asberg M, Bertilsson L, et al: Clomipramine treatment of obsessive disorder: biochemical and clinical aspects. *Arch Gen Psychiatry* 37:1289–1294, 1980.

Traskman L, Asberg M, Bertilsson L, et al: Monoamine metabolites in CSF and suicidal behavior. *Arch Gen Psychiatry* 38:631–636, 1981.

van Kammen DP: 5-HT, a neurotransmitter for all seasons? *Biol Psychiatry* 22:1–3, 1987.

van Praag HM: CSF 5-HIAA and suicide in nondepressed schizophrenics. *Lancet* 2:977–978, 1983.

van Praag HM: Depression, suicide and the metabolism of serotonin in the brain. *J Affective Disord* 4:275–290, 1982.

van Praag HM, Plutchik R, Conte H: The serotonin hypothesis of (auto)aggression: critical appraisal of the evidence, in *Psychobiology of Suicidal Behavior*. Edited by Mann JJ, Stanley M. New York, New York Academy of Sciences, 1986.

Virkkunen M, Nuutila A, Goodwin FK, et al: Cerebrospinal fluid metabolite levels in male arsonists. *Arch Gen Psychiatry* 44:241–247, 1987.

von Andics M: *Über Sinn und Sinnlosigkeit des Lebens*. Vienna, Gerold and Company, 1938.

von Knorring L, Oreland L, Winblad B: Personality traits related to monoamine oxidase activity in platelets. *Psychiatry Res* 12:11–26, 1984.

West DJ: *Murder Followed by Suicide*. London, Heinemann, 1965.

Woolley DW, Shaw E: A biochemical and pharmacological suggestion about certain mental disorders. *Proc Natl Acad Sci USA* 40:228–231, 1954.

Zohar J, Insel TR, Zohar-Kadouch RC, et al: Serotonergic responsivity in obsessive-compulsive disorder: effects of chronic clomipramine treatment. *Arch Gen Psychiatry* 45:167–172, 1988.

6

Serotonin and Violent Behavior

Markku Linnoila
et al.

Abstract: Some of the most striking and controversial implications of serotonin's effects on behavior concern criminal law. In a series of important research papers, Markku Linnoila and his colleagues at the National Institute of Alcohol Abuse and Alcoholism (NIAAA) have explored how low levels of serotonergic activity, when combined with disturbance of glucose metabolism, are found in individuals guilty of arson and impulsive homicide. These impulsive acts are particularly likely when the affected individuals are under the influence of alcohol or drugs. More surprising are the findings that these biochemical deficiencies can be used to predict recidivism in individuals previously convicted of impulsive violence. Because these findings are certain to be of broad interest and importance, we have chosen to publish, in original form, three of the papers in which Dr. Linnoila and his colleagues at the NIAAA have presented their research. It is all the more important to do so, for scholars and lawyers will increasingly need to follow and understand research findings in neuroscience when they are presented in their original scientific form.

Low Cerebrospinal Fluid 5-Hydroxyindoleacetic Acid Concentration Differentiates Impulsive from Nonimpulsive Violent Behavior

M. Linnoila, M. Virkkunen, M. Scheinin, A. Nuutila,
R. Rimon, F. K. Goodwin

Abstract: Relationships of impulsive and nonimpulsive, violent behavior to cerebrospinal fluid (CSF) monoamines and their metabolic concentrations were studied in 36 violent offenders. A relatively low 5-hydroxyindoleacetic acid (5-HIAA) concentration was found in the CSF of impulsive, violent offenders. This was not true for the offenders who had premeditated their acts. Other CSF monoamine or metabolite concentrations were not significantly different between the two groups. Of the groups studied, impulsive, violent offenders who had attempted suicide had the lowest 5-HIAA levels. A low CSF 5-HIAA concentration may be a marker of impulsivity rather than of violence.

In 1976, Asberg et al.[1,2] published two articles concerning low concentrations of 5-hydroxyindoleacetic acid (5-HIAA, the major metabolite of serotonin) in the cerebrospinal fluid (CSF) of a subgroup of depressed patients. The authors remarked that their finding provided partial support to the theory postulating defective serotonergic neurotransmission in depression.[3-5] Furthermore, the Swedish investigators pointed out that the depressed patients with low CSF 5-HIAA concentrations were more likely than depressed patients with normal 5-HIAA to have attempted or completed suicide using particularly violent means. Since their initial reports, Asberg et al. have increased the size of their sample, and the original findings have held up.[6,7] Moreover, Brown et al.[8,9] have demonstrated that patients with character disorders, who are prone to impulsive, violent behavior towards others, have lower concentrations of 5-HIAA in their CSF than do matched controls with character disorders without incidents of impulsive, violent behavior. This finding is consistent with pharmacological models of "aggression" in rodents, where alterations of brain serotonin functions have been linked with increased "aggression."[10,11] A genetic basis for increased violent behavior and reduced CSF 5-HIAA accumulation after probenecid loading in humans has been suggested by two French pilot studies tying this biochemical marker to violent male criminals with the XYY genotype.[12,13]

Another association between a relatively low 5-HIAA concentration in the CSF and abnormal behavior has been demonstrated in alcoholics and their depressed relatives.[14,15] This observation is of particular interest because at least one form of alcoholism in males is associated with criminal and, at times, violent behavior.[16]

We investigated the relationship of both impulsive, violent behavior and premeditated violence to 5-HIAA concentration in the CSF of murderers and attempted murderers undergoing an intensive forensic psychiatric evaluation in the Department of Psychiatry at the University of Helsinki. According to the nature of their acts, subjects were grouped into those who had killed or attempted to kill impulsively (i.e., without provocation or premeditation) a person not close to them and those who had killed or attempted to kill after some premeditation (i.e., the victim was known to the offender and a rationale for the act could be construed in the psychiatric examination). Our hypothesis was that, of the two equally violent groups, the impulsive offenders would have low CSF 5-HIAA concentrations relative to the nonimpulsive offenders who had premeditated their acts. Thus, a low 5-HIAA concentration in the CSF of violent offenders would be more a marker of impulsivity than of violence per se.

METHODS

A total of 36 violent offenders, who signed an informed consent, participated in the study. Participation was voluntary, and the subjects were told that their CSF samples would be used to investigate associations between brain biochemistry and violent behavior. All participants were men, of whom 21 had killed and 15 had attempted to kill their victims. All had used cold weapons, and a common denominator was the unusual cruelty of the index act. Twenty were diagnosed as having intermittent-explosive disorder; seven, antisocial-personality disorder (these were all impulsive offenders); and nine, paranoid or passive-aggressive personality disorder. (These 9 were all offenders who had somehow premeditated their crimes. For the purposes of the present study, they were called *nonimpulsive*.) Furthermore, all subjects fulfilled the criteria for alcohol abuse, and all subjects with intermittent-explosive or antisocial-personality disorder met the criteria for borderline-personality disorder as well. None fulfilled criteria for a schizophrenic or major affective disorder,[17] even though 17 (all in the group of impulsive offenders) had a history of impulsive suicide attempts. Careful history revealed that all impulsive, violent offenders had exhibited disturbed behavior at school compatible with either attention-deficit or aggressive-conduct disorder[17] and had started to abuse alcohol in their early teens. This was not true of the nonimpulsive, violent offenders. According to relatives and court documents, all subjects with explosive- or antisocial-personality disorder had repeatedly demonstrated violent behavior towards others, particularly under the influence of alcohol. Chromosomal analysis conducted on all subjects revealed an XYY genotype in one murderer with intermittent-explosive disorder.

The average age of the subjects was 31.5 ± 8.5 (SD) years, and the average height was 173.6 ± 5.89 cm, with ranges of 18 to 49 years and 160 to 185 cm, respectively.

These demographic variables were similar in the three diagnostic groups. All subjects underwent a one- to two-month intensive forensic psychiatric evaluation and follow-up during which their medication was kept to a minimum and they had no access to alcoholic beverages. Except oxazepam 15 mg for severe insomnia in two subjects, no drugs were given during the last two weeks prior to the lumbar puncture (LP). The patients were maintained on a controlled low-monoamine diet, and they had eight hours of supervised bed rest prior to the LP, which was done through the fourth lumbar space in lateral decubitus position between 8 and 9 A.M. The fluid was collected into polypropylene tubes on ice. Twelve cc of CSF was obtained with an atraumatic lumbar puncture, mixed well and stored in two aliquots at −80°C until analyzed for norepinephrine, 3-methoxy-4-hydroxyphenlglycol (MHPG), dihydroxy-phenylacetic acid (DOPAC), homovanillic acid (HVA), and 5-HIAA with liquid chromatography.[18–20] Norepinephrine and DOPAC were not measured in seven samples because the method was not available when the first samples arrived. All measurements in the laboratory were conducted without knowing the diagnostic grouping of the individuals.

The data were analyzed using an analysis of variance and Student's t-test for independent samples to test the differences between the means of the concentration of each monoamine or metabolite in the three diagnostic groups of violent offenders. Two-tailed probabilities were used.

RESULTS

The difference in the mean (± SD) CSF 5-HIAA concentrations between murderers and attempted murderers was not significant (81.9 ± 21.4 and 72.7 ± 23.9 nM, respectively). Impulsive, violent offenders had significantly lower CSF 5-HIAA concentrations than did paranoid or passive-aggressive (i.e., nonimpulsive) violent offenders (see figure 6.1). The 17 offenders (14 impulsive and three nonimpulsive) who had committed more than one violent crime had a mean (± SD) CSF 5-HIAA concentration of 67.9 ± 12 nM. This was significantly below the CSF 5-HIAA concentration of those offenders who had committed only one violent crime (87.1 ± 23.7; $P < .02$). Moreover, the mean (± SD) CSF 5-HIAA concentration in the impulsive offenders with suicide attempts was 67.4 ± 19.7 nM, which is significantly less than the 91.2 ± 22.0 nM ($p < .01$) found in violent offenders without suicide attempts.

Violent offenders with antisocial personality had lower CSF HVA concentrations than did violent offenders with paranoid or passive-aggressive personality (table 6.1). There were no significant differences in the CSF norepinephrine or other monoamine metabolite concentrations between the diagnostic groups. In the whole sample, the correlation between age and CSF 5-HIAA concentration was not statistically significant ($r = -.02$), whereas height had a negative concentration with CSF 5-HIAA concentration ($r = -.52$; $P < .01$). Age and height were both normally distributed within the diagnostic groups.

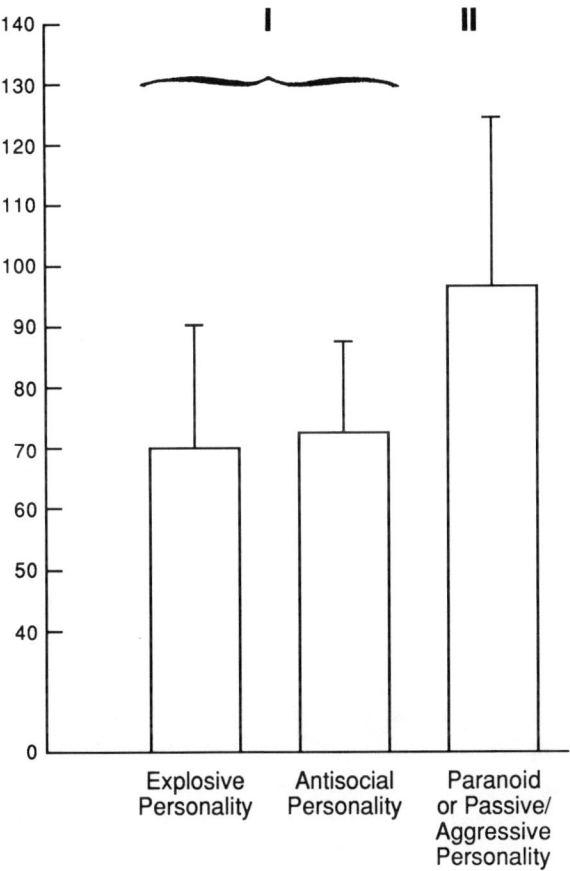

CSF 5-HIAA CONCENTRATIONS
IN VIOLENT OFFENDERS

Pmol/ml

Figure 6.1. CSF 5-HIAA concentration in three diagnostic subgroups of violent offenders. I = impulsive and II = nonimpulsive offenders. ANOVA $P<.01$ between the groups. ** = $P<.05$
Student's t-test, two-tailed, between the nonimpulsive and the
two subgroups of impulsive offenders.

DISCUSSION

A low CSF 5-HIAA concentration seems to be associated with a tendency towards
repeated, impulsive, violent behavior, which can be directed both towards oneself
(suicide attempts) and towards others (murder attempts), and with an early onset of
alcohol abuse.

Our findings do not contradict the earlier suggestion that some depressed patients

Table 6.1. CSF Monoamine and Their Metabolite Concentrations in the Three Diagnostic Groups of Violent Offenders

	Norepinephrine	MHPG	DOPAC	HVA
	nM			
I. Explosive personality	1.1 ± 0.6 (n=16)	31.0 ± 7.2 (n=20)	2.9 ± 1.5 (n=16)	217 ± 97 (n=20)
II. Antisocial personality	1.2 ± 0.4 (n=7)	27.1 ± 6.7 (n=7)	3.3 ± 0.9 (n=7)	184 ± 81* (n=9)
III. Paranoid or passive-aggressive personality	1.5 ± 0.8 (n=7)	36.4 ± 15.4 (n=7)	4.4 ± 2.2 (n=7)	253 ± 46 (n=9)

Source: Reprinted by permission from *Life Sciences*, vol. 33 (1983), pp. 2609–2614, published by Pergamon Press Ltd.

Note: I=impulsive and II=nonimpulsive. *=$P<.05$ compared to respective concentrations in paranoid and passive-aggressive murderers.

without impulsive suicidality could have a relatively low CSF 5-HIAA concentration.[2] The low 5-HIAA subgroup of depressed patients in the Swedish studies may, however, have included patients whose low CSF 5-HIAA concentration could have been produced by withdrawal from previous antidepressant treatment. Thus, the group of impulsively suicidal, depressed patients with low CSF 5-HIAA might have been artificially diluted because both previously drug-free and drug-treated patients were included after a relatively short placebo period. If this reasoning is correct, then the association between low CSF 5-HIAA concentration and impulsive suicidality may be stronger than suggested in previous reports. Prospective studies are needed to answer this question.

A careful inspection of our data reveals that all acid metabolites of monoamines in the CSF are lower in the impulsive than in the paranoid or passive-aggressive murderers, even though the difference in 5-HIAA is the only one that is clearly statistically significant. This complicates the interpretation of the data because all acid metabolites of monoamines are transported out of the CSF by the same active mechanism.[21-24] Thus, a low concentration of 5-HIAA could be due to either a low production rate (a defect in serotonin metabolism) or an increased activity of the transport mechanisms (a membrane phenomenon). This would presume that 5-HIAA concentration is a more sensitive indicator of the activity of the transport mechanism than the other acids. We tend to interpret the data according to the former mechanism, partially because certain animal models of increased "aggression" may be associated with reduced brain serotonin turnover.[10,11]

In another line of investigation, one of the authors (Virkkunen) has found that the impulsive, violent offenders often develop prolonged hypoglycemia due to enhanced insulin secretion after an oral glucose load.[25-27] Furthermore, the hypoglycemia can be aggravated by alcohol. Obviously, interrelationships between low CSF 5-HIAA concentration, drinking history, and regulation of glucose metabolism need to be further elucidated.

Violent offenders with antisocial-personality disorder also had a lower mean CSF HVA concentration than offenders with paranoid or passive-aggressive personality disorder. Any causative significance of the low CSF HVA concentration in patients with antisocial-personality disorder awaits further study.

The implications of our data are limited at the present time because the sample is relatively small and represents an extreme group. However, there may be some persons, in whom a defect in central serotonin metabolism is present, who start to abuse alcohol in their early teens and later become violent psychopaths. In such persons, experimental treatment with serotonergic drugs should be initiated. This strategy could be used initially in subjects who have a family history of the behavioral disorder.[16]

REFERENCES

1. M. Asberg, P. Thoren, L. Traskman, L. Bertilsson, and V. Ringberger, Science 191: 478–480 (1976).
2. M. Asberg, L. Traskman, and P. Thoren, Arch Gen Psychiatry 33: 1193–1197 (1976).
3. G. W. Ashcroft and D. F. Sharman, Nature 186: 1050–1051 (1960).
4. S. J. Dencker, U. Malm, B.-E. Roos, and B. Werdinius, J Neurochem 13: 1545–1548 (1966).
5. A. Coppen, Br J Psychiatry 113: 1237–1264 (1967).
6. M. Asberg, L. Bertilsson, P. Thoren, and L. Traskman, in *Depressive Disorders* (S. Garattini, ed.), pp. 293–305, F. K. Schattauer Verlag, Stuttgart (1978).
7. M. Asberg, L. Bertilsson, E. Rydin, D. Schalling, P. Thoren, and L. Traskman-Benz, Advances of the Biosciences 31: 257–271 (1981).
8. G. L. Brown, F. K. Goodwin, J. C. Ballenger, P. F. Goyer, and L. F. Major, Psychiatry Res 1: 131–139 (1979).
9. G. L. Brown, M. H. Ebert, P. F. Goyer, D. C. Jimerson, W. J. Klein, W. E. Bunney, and F. K. Goodwin, Am J Psychiatry 139: 741–746 (1982).
10. L. Valzelli, Adv Biochem Psychopharmacol 11: 255–263 (1974).
11. J. B. Malick, Curr Dev Psychopharmacol 5: 1–27 (1979).
12. B. Bioulac, M. Benezech, B. Renaud, D. Roche, and B. Noel, Neuropsychobiology 4: 366–370 (1978).
13. B. Bioulac, M. Benezech, B. Renaud, B. Noel, and D. Roche, Biol Psychiatry 15: 917–923 (1980).
14. J. C. Ballenger, F. K. Goodwin, L. F. Major, and G. L. Brown, Arch Gen Psychiatry 36: 224–227 (1979).
15. N. E. Rosenthal, Y. Davenport, R. W. Cowdry, M. H. Webster, and F. K. Goodwin, Psychiatry Res 2: 113–119 (1980).
16. C. R. Cloninger, M. Bohman, and S. Sigvarsson, Arch Gen Psychiatry 38: 861–868 (1981).
17. American Psychiatric Association, *Diagnostic and Statistical Manual III*, Washington, D.C. (1980).
18. B. Petruccelli, G. Bakris, T. Miller, E. R. Korpi, and M. Linnoila, Acta Pharmacol Toxicol 51: 421–427 (1982).
19. M. Scheinin, W.-H. Chang, K. Kirk, and M. Linnoila, Anal Biochem 131: 246–253 (1983).

20. T. Seppala, M. Scheinin, A. Capone, and M. Linnoila, Acta Pharmacol & Toxicol 55: 81–87 (1984).
21. T. L. Sourkes, J Neur Transm 34: 153–157 (1973).
22. R. Sjostrom, J. Eckstedt, and E. Anggard, J Neurol Neurosurg Psychiatry 38: 666–668 (1975).
23. J. A. Kessler, C. S. Patlak, and J. D. Fenstermacher, Brain Res 116: 471–483 (1976).
24. M. Bulat, Brain Res 122: 388–391 (1977).
25. M. Virkkunen and M. O. Huttunen, Neuropsychobiology 8: 30–34 (1982).
26. M. Virkkunen, Neuropsychobiology 8: 34–40 (1982).
27. M. Virkkunen, Br J Psychiatry 142: 598–604 (1983).

Cerebrospinal Fluid Monoamine Metabolite Levels in Male Arsonists

Matti Virkkunen, Arto Nuutila,
Frederick K. Goodwin,
Markku Linnoila

Abstract: Cerebrospinal fluid (CSF) monoamine metabolite levels were studied in 20 arsonists, 20 habitually violent offenders, and 10 healthy inpatient volunteers. The arsonists and violent offenders had been in prison an average of six months before the study. Both the raw data and the data adjusted by analysis of covariance for group differences in age, height, sex, and season of the lumbar puncture showed significantly lower 5-hydroxyindoleacetic acid (5-HIAA) in the arsonists than in other groups. The findings remained the same when arsonists with violent suicide attempts were excluded from the analysis. Although CSF concentrations of 3-methoxy-4-hydroxyphenylglycol (MHPG) or 5-HIAA did not correlate with the severity of repeated fire-setting behavior, low-blood-glucose nadir in the oral glucose tolerance rest (a measure of the tendency toward hypoglycemia) did. These results support the hypothesis that poor impulse control is associated with low levels of certain monoamine metabolites, with a hypoglycemic tendency in criminal offenders.

$\mathbf{S}$everal series of recent clinical studies have provided evidence of an association between indexes of low serotonin turnover in the central nervous system (CNS) and such behaviors as attempted and completed suicide[1-14] and homicide.[15] Furthermore, an indirect index of CNS serotonin turnover, cerebrospinal fluid (CSF) 5-hydroxyindoleacetic acid (5-HIAA) concentration, has been reported to be low in alcoholics[16] and in depressed relatives of alcoholics[17] who themselves are not alcoholic.

Another biochemical variable, which has been previously associated with impulsive fire-setting and violent behaviors,[18,19] as well as suicide attempts in depressed patients,[20] is a low-blood-glucose concentration during the oral glucose tolerance test (GTT). This hypoglycemic response to an oral glucose load seems to be associated with an enhanced insulin response.[21,22] To our knowledge, associations between indexes of a low CNS serotonin (5-hydroxytryptamine [5-HT]) turnover and hypoglycemia, two of the biochemical variables that may be common to individuals exhibiting fire-setting, suicidal, and/or violent behavior, have not been previously investigated.

This essay was originally published in *Archives of General Psychiatry*, Vol. 44 (1987), pp. 241–247. Copyright 1987 American Medical Association. Reprinted with permission. In the public domain.

Despite the relatively numerous reports documenting associations between a low CSF 5-HIAA concentration and several kinds of violent behavior, the exact nature of interactions between this biochemical variable and behavior is obscure. A low CSF 5-HIAA level has been found by various investigators to be associated with acts characterized by violence,[1] aggressiveness,[23] impulsivity,[15] and emotional turmoil.[24,25] To elucidate the specificity of the association between a low CSF 5-HIAA concentration and impulsivity, an association that we have previously emphasized in our report on violent offenders,[15] we studied impulsive arsonists. Behaviorally, these patients represent an extreme of impulsivity and are relatively free of interpersonal aggressiveness and violence. Thus, they allow direct testing of the hypothesis that a low CSF 5-HIAA concentration is primarily associated with impulsivity rather than with violence or aggressiveness.

Arson is an important economic and public-health concern in its own right, costing the United States billions of dollars and an estimated 1000 deaths and 10,000 injuries annually.[26,27] In spite of the magnitude of the problem, there have been few controlled, psychobiologic studies of arsonists. "Pyromania," the classic diagnosis of compulsive arsonists, is included in the *Diagnostic and Statistical Manual of Mental Disorders,*[34] third edition (*DSM-III*) under "Disorders of Impulse Control Not Elsewhere Classified."[28] Studies of various kinds of arson have emphasized the impulsive nature of these acts.[27,29-43] Acts of arson have often been found to be committed by underassertive[43] or schizoid[33] persons with difficulties externalizing their aggressions, choosing as targets inanimate objects rather than persons.[33,39,43] Suicidal tendencies have also been found to be common among arsonists.[29,32,34,44] In a study of 1145 compulsive arsonists who were repeated offenders, Lewis and Yarnell[29] emphasized the central role of insufficient inhibitory control of behavior in these individuals. Only rarely had their patients shown unprovoked, aggressive behavior. Approximately 70% of the adult arsonists in the Lewis and Yarnell[29] series had below-normal intelligence. Besides pyromania, "repeated arsonistic behavior with a lack of motivation,"[29,45] and premeditated arson for profit,[29,45,46] revenge also is a recognized motive for arson.*

In this report, we compare CSF monoamine metabolite concentrations in 20 men who had committed impulsive arson with levels in 20 violent offenders and 10 healthy volunteers. The arsonists also underwent an oral GTT in addition to the lumbar puncture (LP) and careful psychiatric evaluation.

SUBJECTS AND METHODS

Subjects

The patients were 20 nonretarded, male arsonists (IQ > 68) from the Forensic Psychiatry Department of the Helsinki University Central Hospital (group A). Over a period of 23 months, during 1983 and 1984, all 29 successive arsonists from the department, who fulfilled the above criteria, were asked to participate in the study. In Finland, practically all arsonists, except those whose crimes are deliberate insurance frauds, are sent for forensic psychiatric examination. Our sample consisted of adult arsonists who apparently did not set fires for economic gains. Participation was voluntary, and the results of the CSF study did not have any effect on the outcome of the forensic examination and were not available to the court.

The participation of the male violent offenders (group B), used as one of the comparison groups, was secured with the same methods and principles during 14 months of the 1981–1982 time period. Their results have been reported earlier by Linnoila et al.[15] A subgroup of them (n = 20) were matched to the arsonists as closely as possible for age, height, and sex. Both arsonists and violent control offenders had been incarcerated in high-security prisons for an average of six months. They were thus presumably without alcohol and illicit drugs, which could affect CSF monoamine metabolite concentrations, for a prolonged period before the examinations. They were given a low-monoamine diet for a week before the LP. No drugs, except 15 to 50 mg oxazepam for severe insomnia in three arsonists and two violent offenders, were given during the last two weeks before the LP. Even these patients were free of medications the last two days before the LP. All subjects had eight hours of supervised bed rest before the LP. The arsonists and violent offenders were given psychiatric diagnoses using the *DSM-III* criteria[28] at the end of a four- to eight-week forensic psychiatric examination. The criminal offenders also underwent chromosome analysis to detect Klinefelter's syndrome (XYY chromosomal abnormality).

Ten volunteer subjects were studied as healthy controls (group C). They were recruited through the National Institutes of Health, Bethesda, Md, Normal Volunteer Office or by advertisements. All volunteers gave a written, informed consent to participate in the study. They underwent a complete physical examination, including blood and urine analyses, as well as an interview utilizing the Schedule for Affective Disorders and Schizophrenia, Lifetime Version.[49] Only those with lifetime diagnosis of no mental illness were included. The volunteers (three female, seven male) were admitted to a National Institute of Mental Health research ward on the evening before the procedure. All subjects remained at complete bed rest until 9 to 9:30 A.M. the next day, when the LP was performed. The subjects received a low-monoamine diet before the LP for a minimum of 72 hours.

Biochemical Procedure

Lumbar punctures were performed with subjects in the left lateral decubitus position between 8:30 and 9:30 A.M., and 30 mL of CSF was obtained from all subjects. The first 12 mL of CSF was collected on ice, thoroughly mixed, and stored in 1-mL portions at −60°C to be assayed for monoamine metabolites. All samples were analyzed in the same laboratory without knowing the diagnostic groupings of the individuals. Concentrations of 3-methoxy-4-hydroxyphenylglycol (MHPG), 5-HIAA, and homovanillic acid (HVA) were quantified with high-performance liquid chromatography using electrochemical detection.[50]

Oral GTT

Within a few days of the LP, an oral GTT was administered to the arsonists. This is currently a routine procedure in the University of Helsinki Forensic Psychiatry

Department. The tests were preceded by an overnight fast of at least 12 but no more than 16 hours. Stressful examinations were avoided during the two preceding days and the night before. On the morning of the examination, after the fasting blood samples had been taken, the participants were given glucose (Glycodyn), 1 g/kg (4 mL of fluid per kilogram of body weight), which was taken orally as quickly as possible. Blood samples were collected at 30 minutes and at one, two, three, four, and five hours after the ingestion for enzymatic blood-glucose measurements.

Behavioral Measurement

First-degree relatives or significant others filled out a structured questionnaire (M.V., unpublished data, July 1981) concerning the quality and quantity of previous aggressive behavior exhibited by each subject. Detailed police records concerning the index crime and previous criminal records were available as well.

Data Analysis

Analysis of covariance (ANCOVA) was used to assess the effects of diagnostic groups on the dependent variables, CSF 5-HIAA, MHPG, and the HVA concentrations. The covariates were age, height, sex, and season of the LP. They were selected because they have been reported to affect the concentrations of 5-HIAA and HVA in the CSF. These concentrations correlate positively with age and negatively with height[51-55] and exhibit seasonal variations.[54,56,57] Furthermore, they may vary according to the inpatient or outpatient status of the subjects,[58] which was controlled for in the present study by using only inpatients. Post hoc comparisons were made with the Student's t-test.

The univariate (ANCOVA) we used is a special case of multiple-regression analysis designed to answer the question: "Do the groups differ in their mean value of a variable (x_1), allowing for effects of one or more other variables (x_2, x_3, x_n) on the value of x_1?" The effects of x_2 to x_n on x_1 can be "partialled out" using multiple-regression analysis with x_1, as the dependent variable and x_2 to x_n, the covariates, as the independent variables. This regression adjusts the group mean values for x_1 as if each group had the same mean values for the covariates. These adjusted mean values of x_1 are then compared using analysis of variance (ANOVA). The ANCOVA assumes that the effects of the covariates on the dependent variable are the same in each of the groups (i.e., that the slopes of the regression lines do not differ significantly across groups). This requirement was met for all variables whose ANCOVA results are reported below. Two-way ANOVA was used to adjust for the effect of season of LP on metabolite concentrations. Months in which the LPs were performed were divided into the four seasons: March through May as spring, June through August as summer, September through November as fall, and December through February as winter. The seasons were assigned values 1 through 4, starting with spring, for certain of the statistical analyses. In the control group, the two-way ANOVA was used also to adjust for the effect of sex on metabolite concentrations because the control group contained three women. In separate analyses, Pearson product-moment and Spearman rank-order correlations were computed to assess the relationships between monoamine metabolites and blood-glucose nadir in the GTTs. Noncontinuous categorical variables, such as classic motives for arson, and the presence or absence of past aggressive and other criminal behavior were assigned numerical values and entered as dummy variables in the appropriate correlation analyses.

RESULTS

Clinical Characteristics

The age (mean ± SD) of the arsonists was 29.8 ± 9.6 years; the age of the violent offenders was 28.0 ± 9.7 years; and the age of the healthy volunteers was 25.1 ± 3.5 years. The heights were 178 ± 5.0, 174 ± 7.5, and 173 ± 6.5 cm, respectively. Important characteristics of arsonists are given in table 6.2. Nineteen of them fulfilled the *DSM-III* criteria for borderline-personality disorder. One fulfilled the criteria for

Table 6.2. Characteristics of the Arsonists

Patient No.*/ Age, y/No. Arson (During Time)	IQ	5-HIAA nmol/L	HVA nmol/L	MHPG nmol/L	Blood-Glucose** Nadir During GTT mg/dL (nmol/L)	Violent Suicide Attempt(s)***	Arson under Influence of Alcohol
1/24/3 (2 y)	98	38.1	166.4	8.3	34 (1.9)	+	+
2/25/1	69	44.4	227.8	17.8	49 (2.7)	−	+
3/27/1	99	32.2	215.0	17.2	58 (3.2)	−	+
4/30/1	112	27.6	106.5	13.9	54 (3.0)	+	+
5/26/12 (8 y)	97	20.3	121.5	15.8	43 (2.4)	−	+
6/36/1	94	32.9	109.9	13.0	45 (2.5)	+	+
7/47/1	107	79.3	278.3	16.7	43 (2.4)	+	+
8/30/4 (1 h)	80	42.3	271.4	17.3	49 (2.7)	+	+
9/33/2 (1 wk)	105	36.4	152.6	12.9	45 (2.5)	+	+
10/19/1	97	56.6	347.5	26.3	54 (3.0)	−	+
11/44/1	102	24.0	173.7	16.7	63 (3.5)	+	+
12/19/2 (1 y)	115	68.3	358.0	13.9	49 (2.7)	+	+
13/23/3 (1 mo)	82	32.1	181.0	16.6	58 (3.2)	−	−
14/35/1	113	61.5	270.1	15.1	56 (3.1)	+	+
15/31/1	71	28.7	143.5	20.1	43 (2.4)	+	+
16/21/4 (4 mo)	104	42.4	146.0	12.2	56 (3.1)	−	−
17/28/1	88	50.2	218.5	32.0	86 (4.8)	+	+
18/29/2 (1 d)	97	104.9	368.7	14.5	65 (3.6)	−	+
19/17/10 (2 y)	77	88.9	280.2	35.2	40 (2.2)	+	+
20/41/2 (1 wk)	89	28.8	221.1	16.8	43 (2.4)	−	+

*Patient 6 had a major depressive disorder; patient 10 had XYY syndrome; and patient 20 had chronic schizophrenia.

**Values under 54 mg/dL (3.0 mmo/L) are considered hypoglycemic.

***Patient 3 completed suicide after the study.

chronic schizophrenia, one for a major depressive disorder, and one had XYY syndrome. Many of them had occasionally exhibited explosive behavior, usually under the influence of alcohol, and three had committed violent crimes. Twelve had committed crimes against property. Seventeen had symptoms of dysthymic disorder coinciding with their borderline-personality disorder. The depressive symptoms were not, however, of sufficient severity and duration to qualify for the diagnosis of a major depressive disorder. All except three patients also qualified for the *DSM-III* diagnosis of alcohol abuse.

The violent offenders fulfilled the criteria for antisocial personality (n = 5) or intermittent-explosive disorder (n = 15) and a coinciding borderline-personality disorder (n = 20). Seven also had a concomitant dysthymic disorder. They all fulfilled the criteria for alcohol abuse and had a habitual tendency to behave aggressively when under the influence of alcohol. Thirteen had committed a homicide, three had committed homicide attempts, and four had committed grave sexual offenses. Many of them had other kinds of less severe violent crimes and crimes against property in their criminal records.

Results of the ANCOVA

We did not find a significant linear relationship between age and any of the monoamine metabolite concentrations. That is, the correlation coefficients for age and CSF metabolite concentrations among arsonists (5-HIAA, $r = -.07$; MHPG, $r = -.24$; HVA, $r = -.13$) or among violent offenders (5-HIAA, $r = .04$; MHPG, $r = .17$; HVA, $r = -.02$) were not significantly different from 0. The same pertained to height among arsonists (5-HIAA, $r = -.06$; MHPG, $r = .31$; HVA, $r = -.32$) and violent offenders (5-HIAA, $r = -.31$; MHPG, $r = .17$; HVA, $r = -.06$). Season and concentration of HVA had a significant relationship among arsonists (5-HIAA, $r = .39$; MHPG, $r = .16$; HVA, $r = .60$; $P < .01$) but not among violent offenders (5-HIAA, $r = .39$; MHPG, $r = .30$; HVA, $r = -.29$). The HVA concentration in the arsonists was lowest during the spring and rose progressively over the summer and fall to a peak during the winter.

In the ANOVA, age, height, and season were used as covariates. They had no significant effects on 5-HIAA or MHPG concentration (table 6.3).

The effect of these covariates on HVA was different in group A from that for group B (table 6.3) Results of the ANCOVA and the post hoc t-tests, using the adjusted means of the monoamine metabolite concentrations, are given in table 6.4 and figure 6.2. The CSF 5-HIAA concentration was significantly lower in the arsonists than in the violent offenders or controls. The CSF 5-HIAA concentration in the violent offenders was also significantly lower than in the controls. The CSF MHPG concentration was significantly lower in the arsonists when compared with the violent offenders and controls. The CSF MHPG concentration in the violent offenders was lower than in the controls, as well. The CSF HVA concentration did not differ among the groups. When patients with violent suicide attempts were excluded from the analyses, the results remained the same (table 6-4).

Table 6.3. Regression Coefficients

Covariates	5-HIAA	MHPG	HVA
Dependent Variables			
Age	−0.033	0.001	−0.855
	(0.315)	(0.113)	(1.501)
Height	−0.293	0.248	−1.992
	(0.465)	(0.173)	(2.214)
Season	6.232	1.318	9.157
	(2.740)	(0.950)	(13.046)
Analysis of Variance, P			
Equality of adjusted means	.0029	.0000	.8049
Zero slope	.1136	.3089	.6155
Groups Means, nmol/L			
Group means,	47.00	17.62	217.9
A (B)	(69.58)	(28.53)	(222.6)
Adjusted group	49.52	17.34	217.9
means, A (B)	(68.05)	(28.90)	(216.6)

Note: Comparison of arsonists (group A) with violent offenders (group B).

Two-way ANOVA

The two-way ANOVA did not reveal significant effects of group-by-season interactions on CSF 5-HIAA (F = 0.18; *df* = 3, 32), MHPG (F = 1.18; *df* = 3, 27), or HVA (F = 2.81; *df* = 3, 32) concentrations (table 6.3).

Correlations of Monoamine Metabolite Concentrations in Arsonists

5-HIAA—The CSF 5-HIAA concentration in arsonists correlated significantly with CSF HVA (r = .80; n = 20; P < .001) but not with MHPG concentration (r = .31; n = 20). The CSF 5-HIAA concentration correlated negatively with the incidence of criminal acts other than arson among the arsonists (r = −.46; n = 20; P < .05). A low CSF 5-HIAA concentration did not correlate significantly with pyromania as defined in the *DSM-III* (r = −.01; n = 20) or with an established motive of revenge for the index arson (r = −.31; n = 20).

MHPG—The CSF MHPG concentration in arsonists did not correlate with CSF HVA concentration (r = .32; n = 20). The CSF MHPG concentration did not correlate with the incidence of violent criminal acts (r = .18; n = 20) but it correlated positively with crimes, other than arson, against property (r = .43; n = 20; P < .02). Thus, those arsonists with an especially low CSF MHPG level had usually committed only one act of arson or had committed arson exclusively. Similar to a low CSF 5-HIAA concentration, a low MHPG level did not correlate with pyromania according to the *DSM-III* (r = .008; n = 20), but a high MHPG concentration correlated with the motive of revenge (r = 46; n = 20; P < .06).

Table 6.4. ANCOVA Results of Monoamine Metabolites

	Group A*	Group B	Group C	df	F	P	Post Hoc t-Test, P
All Cases							
5-HIAA, nmol/L							
Adjusted means ± SD	48.52 ± 18.19	68.05 ± 16.19	90.00 ± 29.00	2.47	13.832	<.001	A/B, <.002; A/C, <.001; B/C, <.001
n	20	20	10				
MHPG, nmol/L							
Adjusted means ± SD	17.34 ± 6.12	28.90 ± 16.12	42.30 ± 8.50	2.42	47.367	<.001	A/B, <.001; A/C, <.001; B/C, <.001
n	20	15	10				
HVA, nmol/L							
Adjusted means ± SD	223.9 ± 88.63	216.6 ± 86.63	206.0 ± 64.00	2.47	0.1566	.85549	
n	20	20	10				
Reduced Groups							
5-HIAA, nmol/L							
Adjusted means ± SD	46.38 ± 17.37	69.56 ± 17.37	90.00 ± 29.00	2.30	9.724	<.001	A/B, <.005; A/C, <.001; B/C, <.042
n	9	14	10				
MHPG, nmol/L							
Adjusted means ± SD	17.28 ± 5.79	28.18 ± 5.79	42.30 ± 8.50	2.27	32.295	<.001	A/B, <.001; A/C, <.001; B/C, <.001
n	9	11	10				
HVA, nmol/L							
Adjusted means ± SD	216.1 ± 92.32	213.1 ± 92.32	206.0 ± 64.00	2.30	0.03649	.96421	
n	9	14	10				

*ANCOVA indicates analysis of covariance: group A, arsonists; group B, violent offenders; group C, controls. **Reduced groups were groups A and B without patients who had attempted violent suicide.

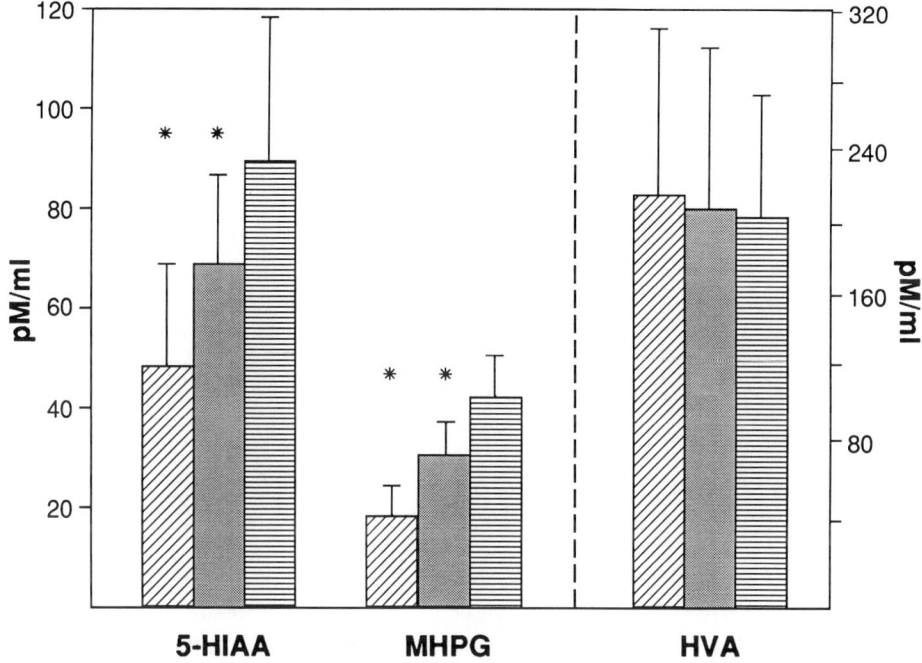

Figure 6.2. Unadjusted mean (± SD) monoamine metabolite concentrations in 20 arsonists (diagonal hatching), 20 violent offenders (cross-hatching), and 10 healthy volunteers (horizontal hatching). Asterisks indicate $P<.001$ compared with controls, using Student's t-test for independent samples.

HVA—The CSF HVA concentration did not correlate with any of the behavioral measurements in arsonists.

Correlations of Monoamine Metabolite Concentrations in Violent Offenders

The CSF 5-HIAA concentration correlated with the CSF HVA concentration in violent offenders, but not as highly as it did among arsonists ($r = .46$; n = 20; $P < .05$). It did not correlate with the CSF MHPG concentration ($r = .07$; n = 20), which did not correlate significantly with the CSF HVA concentration ($r = .04$; n = 20).

Monoamine Metabolites and Blood-glucose Nadir in GTT

There were no significant correlations between the blood-glucose nadir and any of the three monoamine metabolites among these patient groups. The lower limit of normal for the Helsinki University Clinical Chemistry Department is 64 mg/dL (3.0 nmol/L). This is derived as the 95th percentile of a healthy control population. According to this criterion, 11 of the 20 arsonists were hypoglycemic during the GTT. Furthermore, eight arsonists had blood-glucose nadirs of 45 mg/dL (2.5 mmol/L) or less, which indicates a relatively severe reactive hypoglycemia. A low-blood-glucose nadir during the GTT correlated with the incidence of aggressive acts exhibited by the arsonists (r = .57; n = 20; $P <$ .01) and violent offenders (r = .52; n = 20; $P <$ .02) under the influence of alcohol. Different from hypoglycemia, CSF concentrations of any of the monoamine metabolites did not correlate significantly with the same index of aggressive behavior under the influence of alcohol. A history of memory loss under the influence of alcohol correlated positively with a low-blood-glucose nadir among arsonists (r = .66; n = 20; $P <$ .01) and violent offenders (r = .66; n = 20; $P <$ .01). No monoamine metabolites correlated significantly with this aspect of behavior. Pyromania, as defined by the *DSM-III*, correlated with low-blood-glucose nadir during the GTT (r = .51; n = 20; $P <$ .02). The three patients who were the most typical pyromaniacs in the current sample and who had each committed more than one act of arson also had the lowest blood-glucose nadirs among the arsonists (34, 43, and 40 mg/dL [1.9, 2.4, and 2.2 nmol/L]).

COMMENT

The arsonists clearly had the lowest CSF 5-HIAA and MHPG concentrations of the three patient groups we studied. A similar but somewhat weaker finding was true for the violent offenders. Our results concerning comparisons between the healthy volunteers and the offenders have to be considered preliminary because the controls were Americans and the offenders were Finns and because the control group contained three women. Since we did not have permission to perform LPs on healthy volunteers in Helsinki, we chose to use these controls. Furthermore, the controls and offenders were given similar low-monoamine diets prior to the LPs, and in a previous study we found very similar CSF 5-HIAA and HVA concentrations in American and Swedish depressed patients.[58] Finally, we are not aware of any sex differences in the concentrations of these monoamine metabolites in subjects within the age range we have included in the present study.

Earlier studies have not identified any uniform diagnostic groups with a low CSF MHPG concentration, with the possible exception of Korsakoff's psychosis.[59] In a previous study of aggressive military recruits,[23] a weak but positive correlation was found between the mean aggression score and the CSF MHPG level. This finding was not, however, replicated in a later study.[3] In a study by Lidberg et al,[24] CSF MHPG did not differ between homicidal offenders, suicide attempters, and healthy controls.

How strong is the previously reported relationship between a low CSF 5-HIAA concentration and behavior? Low CSF 5-HIAA concentrations have repeatedly been found to be associated with suicidal behavior[1-15,24] and especially with suicide attempts or completed suicides[1-3,6,7,9] using violent means. Furthermore, a low 5-HIAA concentration in the CSF has been found in habitually violent military recruits with a borderline-personality disorder[3,23] and among habitually violent offenders with XYY-syndrome, most of whom had also committed arson.[60] This finding has also been replicated by us among habitually impulsive and violent homicidal offenders with antisocial personality or with an intermittent-explosive disorder.[15] In their Swedish sample, Lidberg et al[24] found especially low CSF 5-HIAA concentrations among homicidal offenders who had killed their sexual partners "in an emotional turmoil" and among suicide attempters who had killed their children.[25] Postmortem brain 5-HT and 5-HIAA concentrations in suicide victims have also been repeatedly found to be low.[61-64] This may be particularly true for the hypothalamus.[65] Furthermore, Stanley et al[66] have reported reduced presynaptic, tritiated, imipramine binding sites, compatible with reduced 5-HT turnover, in frontal-cortex homogenates. Stanley and Mann[67] also found increased postsynaptic 5-HT receptor binding in postmortem brain tissue from suicide victims. These findings concerning 5-HT synapses are in accordance with the low CSF 5-HIAA findings. It seems that low CSF 5-HIAA in violent suicide attempters is relatively independent of the psychiatric diagnosis and possibly related to an as yet ill-defined personality variable or level of stress. In addition to depressions,[1,2,4,7,9,11] this association has been found in schizophrenia,[12,14] in personality disorders, especially the borderline-personality disorder,[3] and in adjustment disorder.[7] The individuals with a violent, antisocial-personality disorder or an intermittent-explosive disorder in our earlier study[15] also fulfilled the diagnostic criteria for borderline-personality disorder. Patients with a bipolar affective disorder seem to be the only major diagnostic group clearly devoid of this association,[5,68,69] and its pertinence to schizophrenic suicides has been recently questioned.[70]

Van Praag and Plutchik,[71] however, are of the opinion that violent suicide attempts in depressed patients are connected with a particularly severe illness, which they call "vital depression."[72,73] Although 80% of the arsonists and violent offenders we studied fulfilled the *DSM-III* criteria for dysthymic disorder, none had an illness of sufficient severity and, especially, duration to qualify for a major depressive disorder or to fit the van Praag and Plutchik description of vital depression. According to a recent study, mood disorders are only rarely associated with serious and repetitive delinquent behavior.[74] When present in criminals, the mood disorders are often secondary to substance abuse or alcoholism, and the depressions tend to be of the agitated type.[74]

There are several possibilities of how a presumed low CNS 5-HT turnover, as indicated by a low CSF 5-HIAA concentration, may be associated with violent behavior: (1) it is associated with a particular depressive syndrome characterized by violent suicide attempts;[71] (2) it is associated with aggression dyscontrol by being conducive of a heightened aggressive drive; (3) it is associated more specifically with a deficiency of impulse control[15] and because of this with dyscontrol of intrapersonal and interpersonal aggression. Patients with affective disorders who have a low CSF 5-HIAA level score highly on indicators of anxiety and hostility in the Rorschach test.[75] A strong negative correlation has also been found between low CSF 5-HIAA concentrations and the psychopathic-deviate scale of the Minnesota Multiphasic Personality Inventory.[3] The results of this study best fit the last hypothesis. Aggressivity and impulse-control problems are, of course, closely related, especially in violent offenders. Impulsive fire-setting behavior is an extreme example of an impulse-control disorder. Because repeated arsonists are particularly impulsive and generally not violent, their very low CSF 5-HIAA concentrations are compatible with the hypothesis of low CNS 5-HT turnover being primarily associated with poor impulse control in humans.

In controlled clinical studies, both tryptophan and lithium carbonate have been found effective in reducing violent acts by habitually impulsive and violent adult offenders[76,77] and by adolescents with undersocialized, aggressive-conduct disorders characterized by severe aggressiveness and explosive impulsivity.[78,79] Sheard and Marini,[80] examining case reports of habitually impulsive and violent offenders who had received lithium carbonate, pointed out "that the anger once aroused does not escalate in the way it customarily does." Shader et al[81] elaborated: "it is as if a delay factor is added by means of lithium as a way-station in patients who previously went automatically from stimulus to response." Children with aggressive-conduct disorder in the study by Campbell et al[79] themselves felt that lithium carbonate "helped to control" them. Thus, it seems that lithium carbonate may rather specifically ameliorate impulse-control problems. In animal studies, both lithium carbonate and tryptophan are effective in reducing mouse-killing "aggression" in the rat.[82] Moreover, lithium carbonate facilitates serotonergic neuronal activity,[83,84] inhibits 5-HT uptake,[85,86] and down-regulates the 5-HT receptor number in a rat's brain.[87,88] Whether lithium carbonate works in arsonists has not, to our knowledge, been studied.

An interesting problem is how abnormalities of CNS monoamine functions, especially 5-HT metabolism, and glucose and insulin metabolism as demonstrated by us are associated with each other in arsonists and habitually violent and impulsive offenders. We did not find significant correlations between the blood-glucose nadir during the GTT and CSF monoamine metabolite concentrations, but low MHPG, 5-HIAA, and blood-glucose nadir coexisted in most arsonists. A tendency for pyromania, repeated arsonist behavior without motive, did not correlate with low monoamine metabolite levels, but it correlated with low-blood-glucose nadir during the GTT. It is known that recidivism after a conviction for arson is quite low among most arsonists. For instance, in the Soothill and Pope[89] 20-year cohort study, only three of 67 arsonists were found to have committed new acts of arson after serving their sentences. In our study, three of the 20 arsonists (subjects 1, 5, and 19) had previous convictions for arson and continued setting fires after release from prison. All three also had a very low blood-glucose nadir during the GTT. Thus, a tendency for severe hypoglycemia may be predictive of recidivism in impulsive arsonists. A lack of correlation between the blood-glucose nadir and CSF 5-HIAA concentration in the present study is not surprising. This is because blood-glucose concentration is multifactorially determined, and the responsiveness of only one of the controlling factors, such as insulin, glucagon, growth hormone, or epinephrine,[90] could be related to central 5-HT metabolism.

The last association to be considered is the intriguing fact that all impulsive, violent offenders and a vast majority of the arsonists we studied fulfill the *DSM-III* criteria for alcohol abuse. Even though this coincidence is striking, there is no direct evidence for a causative link at this time. We have, however, formulated the following testable hypothesis: a low CSF 5-HIAA level is associated with impaired impulse control, dysphoria, and intermittent insomnia. Alcohol, which has a reserpinelike effect on the 5-HT system,[91,92] may acutely remedy some of these symptoms by releasing 5-HT but chronically makes them worse by depleting 5-HT. Thus, alcohol abuse in these individuals with a presumably deficient serotonergic system may represent an attempt to self-medicate.[16] Alcohol, however, only makes the situation worse by further impairing impulse control and possibly aggravating hypoglycemia. Such an association could explain why so many of these subjects were drunk when committing their crimes. Supportive of our hypothesis, Branchey et al[93] have found evidence that, among alcoholics with a history of both depression and aggression, the ratio in plasma of tryptophan to large neutral amino acids may be low. Additional studies of alcohol abuse, violent behavior, and glucose and 5-HT metabolism are clearly needed. Controlled experiments to determine effects of serotonergic antidepressants in violent offenders and arsonists should also be initiated.

REFERENCES

1. Asberg M, Traskman L, Thoren P: 5-HIAA in the cerebrospinal fluid: a biochemical suicide predictor? *Arch Gen Psychiatry* 1976;33:1193–1197.
2. Asberg M, Bertilsson L, Martensson B, Scalia-Tomba G-P, Thoren P, Traskman-Bendz L: CSF monoamine metabolites in melancholia. *Acta Psychiatr Scand* 1984;69:201–219.
3. Brown GL, Ebert MH, Goyer PF, Jimerson CD, Klein WJ, Bunney WE, Goodwin FK: Aggression, suicide and serotonin: relationships to CSF amine metabolites. *Am J Psychiatry* 1982;139:741–746.
4. Agren H: Symptom patterns in unipolar and bipolar depression correlating with monoamine metabolites in the cerebrospinal fluid: II. Suicide. *Psychiatry Res* 1980;2:225–236.
5. Agren H: Life at risk: markers of suicidality in depression. *Psychiatr Rev* 1983;1:87–104.
6. Banki CM, Molnar G, Vojnik M: Cerebrospinal fluid amine metabolites, tryptophan and clinical parameters in depression: II. Psychopathological symptoms. *J Affective Disord* 1981;3:91–99.
7. Banki CM, Arato M, Papp Z, Kurcz M: Biochemical markers in suicidal patients: investigations with cerebrospinal fluid amine metabolites and neuroendocrine tests. *J Affective Disord* 1984;6:341–350.
8. Banki CM, Arato M: Amine metabolites, neuroendocrine findings, and personality dimensions as correlates of suicidal behavior. *Psychiatry Res* 1983;10:253–261.
9. Traskman L, Asberg M, Bertilsson L, Sjostrand L: Monoamine metabolites in CSF and suicidal behavior. *Arch Gen Psychiatry* 1981;38:631–636.
10. Oreland L, Wiberg A, Asberg M, Traskman L, Sjostrand L, Thoren P, Bertilsson L, Tybring G: Platelet MAO activity and monoamine metabolites; in cerebrospinal fluid in depressed and suicidal patients and in healthy controls. *Psychiatry Res* 1981;4:21–29.
11. Van Praag HM: Depression, suicide and the metabolites of serotonin in the brain. *J Affective Disord* 1982;4:275–290.
12. Van Praag HM: CSF 5-HIAA and suicide in non-depressed schizophrenics. *Lancet* 1983;2:977–978.
13. Montgomery SA, Montgomery D: Pharmacological prevention of suicidal behavior. *J Affective Disord* 1982;4:291–298.
14. Ninan PT, van Kammen DP, Scheinin M, Linnoila M, Bunney WE, Goodwin FK: CSF 5-hydroxyindoleacetic acid levels in suicidal schizophrenic patients. *Am J Psychiatry* 1984;141:566–569.
15. Linnoila M, Virkkunen U, Scheinin M, Nuutila A, Rimon R, Goodwin FK: Low cerebrospinal fluid 5-hydroxyindoleacetic acid concentration differentiates impulsive from nonimpulsive violent behavior. *Life Sciences* 1983;33:2609–2614.
16. Ballenger JC, Goodwin FK, Major LF, Brown GL: Alcohol and central serotonin metabolism in man. *Arch Gen Psychiatry* 1979;36:224–227.

17. Rosenthal N, Davenport Y, Cowdry R, Webster M, Goodwin F: Monoamine metabolites in cerebrospinal fluid of depressive subgroups. *Psychiatry Res* 1980;2:113–119.
18. Virkkunen M: Reactive hypoglycemic tendency among habitually violent offenders. *Neuropsychobiology* 1982;8:36–40.
19. Virkkunen M, Huttunen MO: Evidence for abnormal glucose tolerance test among violent offenders. *Neuropsychobiology* 1982;8:30–34.
20. Henninger G, Mueller P, Davis L: Progressive symptoms and the glucose tolerance test and insulin tolerance test. *J Nerv Ment Dis* 1975;181:421–432.
21. Virkkunen M: Insulin secretion during the glucose tolerance test in antisocial personality. *Br J Psychiatry* 1983;142:598–604.
22. Virkkunen M: Insulin secretion during the glucose tolerance test among habitually violent and impulsive offenders. *Aggressive Behav* 1986;12:303–310.
23. Brown GL, Goodwin FK, Ballenger JC, Goyer PF, Major LF: Aggression in humans correlates with cerebrospinal fluid amine metabolites. *Psychiatry Res* 1979;1:131–139.
24. Lidberg L, Tuck JR, Asberg M, Scalia-Tomba GP, Bertilsson L: Homicide, suicide and CSF 5-HIAA. *Acta Psychiatr Scand* 1985;71:230–236.
25. Lidberg L, Asberg M, Sundquist-Stensman UB: 5-Hydroxyindoleacetic acid levels in attempted suicides who have killed their children. *Lancet* 1984;2:928.
26. Rider AO: The fire-setter, a psychological profile: II. *Law Enforcement Bull* 1980;49:7–21.
27. Koson DF, Dvoskin J: Arson: a diagnostic study. *Am Acad Psychiatry Law Bull* 1982;10:39–49.
28. American Psychiatric Association Committee on Nomenclature and Statistics: *Diagnostic and Statistical Manual of Mental Disorders*. 3rd ed. Washington, DC: American Psychiatric Association; 1980.
29. Lewis NDC, Yarnell H: *Pathological Fire-setting (Pyromania)*. Nervous and Mental Disease Monographs, 82. New York: The Coolidge Foundation; 1951.
30. McKerracher BW, Dacre AJI: A study of arsonists in a special security hospital. *Br J Psychiatry* 1966;112:1151–1154.
31. Hurley W, Monahan TM: Arson: the criminal and the crime. *Br J Criminol* 1969;9:4–21.
32. Tennent TG, McQuaid A, Loughnane T, Hands AJ: Female arsonists. *Br J Psychiatry* 1971;119:497–502.
33. Virkkunen M: On arson committed by schizophrenics. *Acta Psychiatr Scand* 1974;50:152–160.
34. Virkkunen M: Reactive hypoglycemic tendency among arsonists. *Acta Psychiatr Scand* 1984;69:445–452.
35. Gunderson JG: Management of manic states: the problem of fire setting. *Psychiatry* 1974;37:137–146.
36. Boling L, Brotman C: A fire-setting epidemic in a state mental health center. *Am J Psychiatry* 1975;132:946–950.
37. Cowen P, Mullen PE: An XYY man. *Br J Psychiatry* 1979;135:79–81.
38. Dalton K: Cyclical criminal acts in premenstrual syndrome. *Lancet* 1980;2:1070–1071.
39. Hill RW, Langevin R, Paitich D, Handy L, Russon A, Wilkinson L: Is arson an aggressive act or a property offense? A controlled study of psychiatric referrals. *Can J Psychiatry* 1982;27:648–654.
40. Stewart MA, Culver KW: Children who set fires: the clinical picture and a follow-up. *Br J Psychiatry* 1982;140:357–363.
41. Pagcoe H: Edmonton's arson epidemic: February 1980. *Med Law* 1983;2:173–180.
42. Yesavage JA, Benezech M, Ceccaldi P, Bourgeois M, Addad M: Arson in mentally ill and criminal populations. *J Clin Psychiatry* 1983;44:128–130.

43. Harris GT, Rice ME: Mentally disordered fire-setters. *Int J Law Psychiatry* 1984;7:19–34.
44. Ashton JR, Donnan SPB: Suicide by burning: a current epidemic. *Br Med J* 1979;280:769–770.
45. Mavromatis M, Lion JR: A primer of pyromania. *Dis Nerv Syst* 1977;38:954–955.
46. Molnar G, Keitner L, Harwood BT: A comparison of partner and solo arsonists. *J Forensic Sci* 1984;29:574–583.
47. Robbins E, Robbins L: Arson with special reference to pyromania. *NY State J Med* 1967;67:795–798.
48. Bradford JMW: Arson: a clinical study. *Can J Psychiatry* 1982;27:188–193.
49. Endicott J, Spitzer RL: A diagnostic interview: the schedule for affective disorders and schizophrenia. *Arch Gen Psychiatry* 1978;35:837–844.
50. Scheinin M, Chang W-H, Kirk K, Linnoila M: Simultaneous determination of 3-methoxy-4-hydroxyphenylglycol, 5-hydroxyindoleacetic acid and homovanillic acid in cerebrospinal fluid with high-performance liquid chromatography using electrochemical detection. *Anal Biochem* 1983;131:246–253.
51. Bowers MB, Gerbode FA: Relationship of monoamine metabolites in human cerebrospinal fluid to age. *Nature* 1968;219:1256–1257.
52. Gottfries CG, Gottfries I, Johansson B, Olsson R, Persson T, Roos B-E, Sjostrom R: Acid monoamine metabolites in human cerebrospinal fluid and their relations to age and sex. *Neuropharmacology* 1971;10:665–672.
53. Asberg M, Bertisson L, Thoren P, Traskman L: CSF monoamine metabolites in depressive illness. In Grattini S, ed. *Depressive Disorders*. Stuttgart: Schattauer; 1978:293–305.
54. Wode-Helgodt B, Sedvall G: Correlations between height of subject and concentrations of monoamine metabolites in cerebrospinal fluid from psychotic men and women. *Commun Psychopharmacol* 1978;10:665–672.
55. Losonczy MF, Mohs RC, Davis KL: Seasonal variations of human lumbar CSF neurotransmitter metabolite concentrations. *Psychiatry Res* 1984;12:79–87.
56. Moir ARB, Ashcroft GW, Crawford TBB, Eccleston D, Guldberg HC: Cerebral metabolites in cerebrospinal fluid as a biochemical approach to the brain. *Brain* 1970;93:357–368.
57. Guthrie S, Berrettini W, Rubinow DR, Nurnberger J, Linnoila M: Different neurotransmitter metabolite concentrations in CSF samples from inpatient and outpatient normal volunteers. *Acta Psychiatr Scand* 1986;73:315–321.
58. Agren H, Mefford IN, Rudorfer MV, Linnoila M, Potter WZ: Interacting neurotransmitter systems: a non-experimental approach to the 5HIAA-HVA correlation in human CSF. *Psychiatry Res*, in press.
59. McEntee WJ, Mair RG: Memory impairment in Korsakoff's psychosis: a correlation with brain noradrenergic activity. *Science* 1978;202:905–907.
60. Bioulac B, Benezech M, Renaud B, Noel B, Roche D: Serotoninergic dysfunction in the 47, XYY syndrome. *Biol Psychiatry* 1980;15:917–923.
61. Shaw DM, Camps FE, Eccleston EG: 5-Hydroxytryptamine in the hind-brain of depressive suicides. *Br J Psychiatry* 1967;113:1407–1411.
62. Bourne HR, Bunney WE Jr, Colburn RW, Davis JM: Noradrenaline 5-hydroxytryptamine and 5-hydroxyindoleacetic acid in hind-brains of suicidal patients. *Lancet* 1968;2:805–808.
63. Pare CMB, Yeung DPH, Price K, Stacey RS: 5-hydroxytryptamine, noradrenaline and dopamine in brainstem, hypothalamus and caudate nucleus of controls and of patients committing suicide by coal-gas poisoning. *Lancet* 1969;2:133–135.
64. Lloyd KG, Farley W, Deck JHN, Hornykiewicz O: Serotonin and 5-hydroxyindoleacetic acid in discrete areas of the brainstem of suicide victims and control patients. *Adv Biochem Psychopharmacol* 1974;11:387–398.

65. Korpi E, Kleinman JE, Goodman S, Phillips I, DeLisi L, Linnoila M, Wyatt RJ: Serotonin and 5-hydroxyindoleacetic acid concentrations in different brain regions of suicide victims. *Arch Gen Psychiatry* 1986;43:594–600.

66. Stanley M, Virgilio J, Gershon S: Tritiated imipramine binding sites are decreased in the frontal cortex of suicides. *Science* 1982;216:1337–1339.

67. Stanley M, Mann JJ: Increased serotonin-2 binding sites in frontal cortex of suicide victims. *Lancet* 1983;1:214–216.

68. Roy-Byrne P, Post RM, Rubinow DR, Linnoila M, Savard R, David D: CSF 5-HIAA and personal and family history of suicide in affectively ill patients: a negative study. *Psychiatry Res* 1983;10:263–274.

69. Berrettini WH, Nurnberger JI, Scheinin M, Seppala T, Linnoila M, Narrow W, Simmons-Alling S, Gershon E: Cerebrospinal fluid and plasma monoamines and their metabolites in euthymic bipolar patients. *Biol Psychiatry* 1986;20:257–269.

70. Roy A, Ninan P, Mazonson MA, Pickar D, van Kammen D, Linnoila M, Paul SM: CSF monoamine metabolites in chronic schizophrenic patients who attempt suicide. *Psychol Med* 1985;15:335–340.

71. van Praag HM, Plutchik R: Depression type and depression severity in relation to risk of violent suicide attempt. *Psychiatry Res* 1984;12:333.

72. van Praag HM: *Monoamine Oxidase Inhibition as a Therapeutic Principle in the Treatment of Depression*, Utrecht, the Netherlands: University of Utrecht; 1962. Thesis.

73. van Praag HM: *Psychotropic Drugs: A Guide for the Practitioner.* New York: Brunner/Mazel Inc; 1978.

74. Alessi NE, McManus M, Grapentine WL, Briekman A: The characterization of depressive disorders in serious juvenile offenders. *J Affective Disord* 1984;6:9–17.

75. Rydin E, Schalling D, Asberg M: Rorschach ratings in depressed and suicidal patients with low CSF 5-HIAA. *Psychiatry Res* 1982;7:229–243.

76. Young SN, Chouinard G, Annable L, Morand C, Ervin FR: The therapeutic action of tryptophan in depression, mania and aggression. In: Schlossberger HG, Koehen W, Linzen B, Steinhart H, eds. *Progress in Tryptophan and Serotonin Research.* New York: Walter de Gruyter & Co; 1984:321–324.

77. Sheard MH, Marini JL, Bridges CI, Wagner E: The effect of lithium on impulsive aggressive behavior in man. *Am J Psychiatry* 1976;133:1409.

78. Campbell M, Cohen IL, Small AM: Drugs in aggressive behavior. *J Am Acad Child Psychiatry* 1982;21:107–117.

79. Campbell M, Small AM, Green WH, Jennings SJ, Perry R, Bennett WG, Anderson L: Behavioral efficacy of haloperidol and lithium carbonate. *Arch Gen Psychiatry* 1984;41:650–656.

80. Sheard MH, Marini JL: Treatment of human aggressive behavior: four case studies of the effect of lithium. *Compr Psychiatry* 1978;19:37–45.

81. Shader RI, Jackson AH, Dodes LM: The antiaggressive effects of lithium in man. *Psychopharmacology* 1974;40:17–24.

82. Broderick P, Lynch V: Behavioral and biochemical changes induced by lithium and L-tryptophan in muricidal rats. *Neuropharmacology* 1982;21:671–679.

83. Sangdee C, Franz DN: Lithium enhancement of central 5-HT transmission induced by 5-HT precursors. *Biol Psychiatry* 1980;15:59–75.

84. Knapp S, Mandell AJ: Short- and long-term lithium administration: effects on the brain's serotonergic biosynthetic systems. *Science* 1973;180:645–647.

85. Swann AC, Heninger GR, Roth RH, Maas JW: Differential effects of short- and long-term

lithium on tryptophan uptake and serotonergic function in cat brain. *Life Sciences* 1981;28:347–354.

86. Ahluwalia P, Singhal RL: Monoamine uptake into synaptosomes from various regions of rat brain following lithium administration and withdrawal. *Neuropharmacology* 1981;20:483–487.

87. Treiser S, Kellar KJ: Lithium: effects on serotonin receptors in rat brain. *Eur J Pharmacol* 1980;64:183–185.

88. Maggi A, Enna SJ: Regional alterations in rat brain neurotransmitter systems following chronic lithium treatment. *J Neurochem* 1980;34:888–892.

89. Soothill KL, Pope PJ: Arson: a 20-year cohort study. *Med Sci Law* 1973;13:127–138.

90. Cryer PE, Gerieh JE: Glucose counterregulation, hypoglycemia, and intensive insulin therapy in diabetes mellitus. *N Engl J Med* 1985;313:232–241.

91. Holman RB, Shape BM: Effects of ethanol on 5-hydroxytryptamine release from rat corpus striatum in vivo. *Alcohol* 1985;22:249–253.

92. Goodwin DW: Alcoholism and genetics. *Arch Gen Psychiatry* 1985;42:171–174.

93. Branchey L, Branchey M, Shaw S, Lieber CS: Depression, suicide and aggression in alcoholics and their relationships to plasma amino acids. *Psychiatry Res* 1984;12:219–226.

*Note to Page 70: References 22, 27, 31, 33, 34, 39, 41, 42, 44–48.

Relationship of Psychobiological Variables to Recidivism in Violent Offenders and Impulsive Fire Setters: A Follow-up Study

Matti Virkkunen, Judith DeJong,
John Bartko, Frederick K. Goodwin,
Markku Linnoila

Abstract: Fifty-eight violent offenders and impulsive fire setters were followed for an average of three years after release from prison. Recidivists who committed a new violent offense or act of arson had significantly lower CSF 5-HIAA and HVA concentrations and blood-glucose nadirs after oral glucose challenge than did nonrecidivists. A discriminant analysis, based on blood-glucose nadir and CSF 5-HIAA concentration correctly classified 84.2% of the subjects.

In 1976, Asberg et al[1] published their observation of an association between a low concentration of 5-hydroxyindoleacetic acid (5-HIAA) in the cerebrospinal fluid (CSF) and violent suicide attempts in patients with unipolar depression. This classic study has been replicated by several groups of investigators,[2-5] and the finding has been expanded to patients with other psychiatric diagnoses,[6-8] except to those with bipolar affective disorder.[9-11] The reasons for the apparent difference in biochemical concomitants of suicide attempts between patients with bipolar affective disorder and patients with other mental disorders are unclear.

Brown et al[12] later reported that a low CSF 5-HIAA concentration was associated with outward-directed aggression, as well. Based on studies of violent offenders[13] and impulsive fire setters,[14] we subsequently made the suggestion that a low CSF 5-HIAA concentration was associated primarily with a disorder of impulse control rather than with aggressiveness or violence as such. In the latter study, we also found that, compared to healthy volunteers, eight of 20 impulsive fire setters had both a low CSF 5-HIAA concentration (< 45 nmol/L) and a low-blood-glucose nadir (< 3.0 mmol/L) after an oral glucose challenge.[14] Previously, Virkkunen[15,16] had observed that a low-blood-glucose nadir was associated with poor sleep and a lifetime history of antisocial behaviors.

Recently, Asberg et al[17] and Roy et al[18] have in follow-up studies shown that CSF

This artical was originally published in *Archives of General Psychiatry*, Vol. 46 (July 1989). pp. 600–603.
Copyright 1989 American Medical Association. Reprinted with permission. In the public domain.

5-HIAA or homovanillic acid (HVA) concentrations are relatively powerful predictors of suicide risk. CSF 5-HIAA and HVA concentrations usually are highly correlated with each other. Furthermore, a "concentration-response" relationship exists between CSF 5-HIAA and HVA concentrations, indicating that the HVA concentration is partially determined by the 5-HIAA concentration. Thus, the correlation probably reflects physiologically meaningful serotonin-dopamine interactions in the central nervous system (CNS).[19] Consequently, a low concentration of either compound in the CSF may be an indicator of the same pathophysiological process, even though a low CSF HVA concentration has been associated with suicide attempts only in patients with unipolar depression.[17]

We tested in a follow-up study the degree to which psychobiological variables, which in previous studies had discriminated suicide attempters and impulsive offenders from other patients and controls, could discriminate recidivists from nonrecidivists in a sample of violent offenders and impulsive arsonists. The specific hypothesis was that a low CSF 5-HIAA or HVA concentration and a low-blood-glucose nadir would be associated with recidivism.

METHODS

The subjects were 58 men with a mean age (all variances are expressed as SDs) of 30.2 + 10.0 years and a mean Wechsler Adult Intelligence Scale, full-scale IQ of 96.5 + 15.3. Two subjects had an IQ below 70 (one 68, the other 53). 51 subjects had participated in the two studies reported previously.[14,15] Similar data on seven additional subjects were available for this study.

Characteristics of the Original Crime

Thirty-six offenders had either attempted or committed murder, and 22 were arsonists. Based on the police reports of the original crimes, violent offenders were divided into impulsive (n = 24) and nonimpulsive (n = 12). The offense was categorized by a senior research psychiatrist (M.V.) as impulsive if the victim was previously unknown to the offender and no provocation was evident and, furthermore, if the attack did not represent an attempt to rob the victim. The nonimpulsive offenders knew their victims, and there was evidence of premeditation of the crime. All arsonists were selected to represent impulsive fire setters.[14] Thus, the present sample consisted of 46 impulsive and 12 nonimpulsive subjects.

Psychiatric Diagnoses

Psychiatric diagnoses were made according to the *Diagnostic and Statistical Manual of Mental Disorders*, third edition (*DSM-III*), criteria[20] by a senior research psychiatrist (M.V.) who was not blind to the criminal records and past histories of the patients. Of the nonimpulsive, violent offenders, five fulfilled the criteria for passive-aggressive personality disorder and six for paranoid-personality disorder. Of the impulsive, violent offenders, six had antisocial-personality disorder, and 18 had intermittent-explosive disorder. All violent offenders fulfilled the criteria for alcohol abuse and all impulsive violent offenders for borderline-personality disorder, as well. Eight violent offenders fulfilled diagnostic criteria for dysthymic disorder, and one had episodes of major depression. Seventeen fire setters had borderline-personality disorder, and 15 had intermittent-explosive disorder. Twenty fulfilled the criteria for alcohol abuse and 18 for dysthymic disorder, three with distinct superimposed periods of major depression. The diagnostic characteristics of the subjects are presented in detail in table 6.5.

Follow-up

This report contains data obtained during a mean follow-up period of 35.6 ± 18.0 months after the subjects' release from prison. The criminal register of Finland was searched for repeat crimes at that time. When a registered, new, violent offense or act of arson was found, police reports, court documents, and hospital records were reviewed.

Biochemical Measurements

Samples of CSF were obtained under drug-free conditions when the subjects were maintained on a low-monoamine diet during the initial studies.[13,14] Concentrations of the major metabolites of dopamine, norepinephrine, and serotonin—HVA, 3-methoxy-4-hydroxyphenlglycol (MHPG), and 5-HIAA, respectively—were measured with

Table 6.5. *DSM-III* Diagnoses of 58 Violent Offenders and Impulsive Fire Setters Were Followed Up for Recidivism

	Violent Offenders		
	Impulsive (n=24)	Nonimpulsive (n=12)	Arsonists Impulsive (n=22)
DSM-III Axis I			
Dysthymic disorder	3	5	18
Major depressive episode	0	1	3
Intermittent-explosive disorder	18	0	15
Alcohol abuse	24	12	20
DSM-III Axis II			
Antisocial-personality disorder	6	0	3
Borderline-personality disorder	24	0	17
Paranoid-personality disorder	0	6	0
Passive-aggressive personality disorder	0	5	0
History of suicide attempts	10	1	16

high-performance liquid chromatography using electrochemical detection. The within- and between-run coefficients of variation of the method are less than 10%.[21] All biochemical analyses were performed by personnel unaware of the clinical state of the subjects.

Glucose Tolerance Tests

Oral glucose tolerance tests were administered to all subjects after a 12- to 16-hour overnight fast.[14] At 8 A.M. the subjects received 1 g/kg (4 mL/kg) of body weight of glucose solution, which they ingested as quickly as possible. Blood samples were collected before glucose administration and 30 minutes, one, two, three, four, and five hours after it. Blood-glucose concentrations were determined enzymatically.

DATA ANALYSES

All analyses were computed using the SPSSx statistical package.[22] The biochemical variables, the man's age at the time of follow-up, his IQ, and the number of months since his release from prison were subjected to a Pearson correlation analysis.

Linear discriminant analysis was computed in an attempt to predict membership in the two groups of interest: recidivists (n = 13) and nonrecidivists (n = 45). A cutoff Fin/Fout = 4 was used for entering a variable into the discriminant function. The variables meeting this criterion were blood-glucose nadir and CSF 5-HIAA concentration.

The subjects were classified as recidivists (n = 13) and nonrecidivists (n = 45). After the group variances were examined for homogeneity, t-tests were used to compare the groups.

Two-by-two table chi-squares were computed as the final step to examine the presence of dysthymic disorder and impulsivity as predictors of the recidivist and nonrecidivist outcomes. Yate's correction was used when a cell contained fewer than five subjects.

RESULTS

Biochemical Variables

CSF 5-HIAA concentration was available in all 58 subjects; the mean was 68.9 ± 26.9 nmol/L. Mean CSF HVA was 209.3 ± 83.3 nmol/L (n = 44); and MHPG, 26.5 ± 12.2 nmol/L (n = 44). The mean blood-glucose nadir during the oral glucose tolerance test was 2.92 ± 0.56 mmol/L (n = 57).

Correlation Analysis

CSF 5-HIAA concentration correlated with both CSF HVA (r = .52; P < .001) and MHPG (r = .48; P < .001) concentrations. Blood-glucose nadir showed only weak correlations with CSF monoamine metabolite concentrations (e.g., 5-HIAA vs. blood-glucose nadir, $r = .05$; $P = NS$). The length of time outside prison prior to committing a new crime did not correlate with any of the independent variables.

Linear Discriminant Analysis

Blood-glucose nadir during the oral glucose tolerance test correctly classified 80.7% of the subjects. Adding CSF 5-HIAA concentration, which was the only independent variable other than blood-glucose nadir meeting the inclusion criterion (i.e., improving significantly discrimination between the groups), increased the percentage of subjects correctly classified to 84.2% (table 6.6). Because the number of subjects in the sample falling into the classifications of recidivists (n = 13) and nonrecidivists (n = 44) was disproportionate, the accuracy of the prediction was also disproportionate in the two groups (95.5% for the nonrecidivists, sensitivity; and 46.2% for the recidivists, specificity). The results of the linear discriminant analysis are presented graphically

Table 6.6. Results of Discriminant Analysis to Predict the Outcome of a Repeated Offense

Stepwise Entry	Discriminant Analysis				
Step	Variable	Percentage Correctly Classified*	F	dt	P
1	Blood-glucose nadir	80.7%	15.50	1.56	.0002
2	5-Hydroxyindoleacetic acid	84.2%	10.24	2.55	.0002

Classification Matrix after First Step

Actual Group	Predicted Membership, %		Total N
	No Offense	Offense (N)	
No offense	97.7 (43)	2.3 (1)	44
Offense	76.9 (10)	23.1 (3)	13

Classification Matrix after Second Step

Actual Group	Predicted Membership, %		Total N
	No Offense	Offense (N)	
No offense	95.5 (42)	4.5 (2)	44
Offense	53.8 (7)	46.2 (6)	13

*The total number of subjects classified correctly is 42+6=48; 48/57=84.2%. The expected number of correct classification by chance based on the prior probabilities of the group sizes is: (13/57) (13/57) + (44/57) (44/57) = 0.65; 0.65 × 57 = 37 chance correct classifications. Via the binomial approximation of the z statistic for the statistical difference of 48 actual vs. 37 chance expected is $(48 - 37 = 11 \sqrt{(0.65 \times 0.35 \times 57)} = 3.05, P<.01$.

in figure 6.3. The positive predictive value of the discriminant function (i.e., probability of being a recidivist given the prediction of such by the discriminant function) is $6/8 = 0.75$.

When the CSF 5-HIAA concentration was entered alone, it classified all subjects as nonrecidivist. Consequently, the percentage of correct classifications is 44 recidivists divided by 57, or 77.2%. Thus, 5-HIAA by itself cannot be regarded as a predictor variable for recidivism. However, in the stepwise discriminant-function analysis (table 6.6), where CSF 5-HIAA was the second variable to enter after the blood-glucose nadir, the percentage correctly classified rose to 84.2%, with almost all of the nonrecidivists and slightly less than half of the recidivists correctly classified.

T-tests

Age (recidivists 29.2 ± 7.7, n = 13; and nonrecidivists 30.5 ± 10.7, n = 45) and IQ (recidivists 94.3 ± 18.4; and nonrecidivists 97.2 ± 14.4) did not differ significantly between the groups. On the other hand, the t-tests showed significant differences between the groups for CSF 5-HIAA ($t = 2.31$; $P < .025$) and HVA ($t = 2.47$; $P = .018$) concentrations and blood-glucose nadir during the oral glucose tolerance test ($t = 3.47$; $P = .001$). There was no difference between the two groups for CSF MHPG concentration (figure 6.4).

Chi-squares were not significant.

COMMENT

The present study is the first to examine prospectively the predictive power of selected behavioral and psychobiological variables for recidivism in a sample of violent offenders and arsonists. The strongest predictors are blood-glucose nadir during the oral glucose tolerance test and CSF 5-HIAA concentration. A discriminant function based on blood-glucose nadir and CSF 5-HIAA concentration correctly classified 84.2% of the subjects as to the outcomes of recidivism and nonrecidivism (table 6.6). The psychobiological variables as such or in combination with the behavioral variables had more predictive power for the outcome than did any combination of behavioral variables. This finding agrees with existing literature. According to several recent reports, behavioral and diagnostic variables are not very useful in predicting long-term outcome among dangerous subjects.[23–25]

The major questions about the present data are their validity and generalizability. The psychiatric diagnoses were determined by an investigator who knew the backgrounds of the subjects. Though they were the result of four to six weeks' forensic psychiatric evaluation, blind and structured interviews would have been preferable. Furthermore, the number of repeat offenses among the arsonists was higher, and they committed more violent offenses than expected based on previous studies.[26,27] This was true despite the fact that only newly registered crimes were used to define outcomes. Registered crimes are probably an underestimate of new crimes committed.

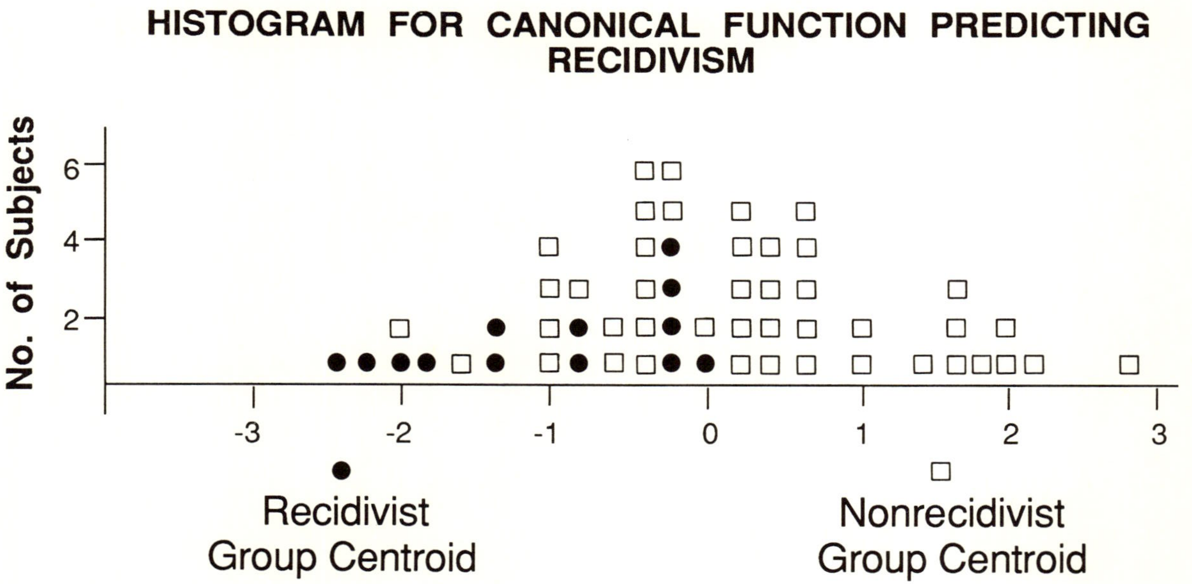

Figure 6.3. Graphic presentation of results of the discriminant analysis to predict outcome. $N = 57$ because the oral glucose tolerance test was not available for one subject.

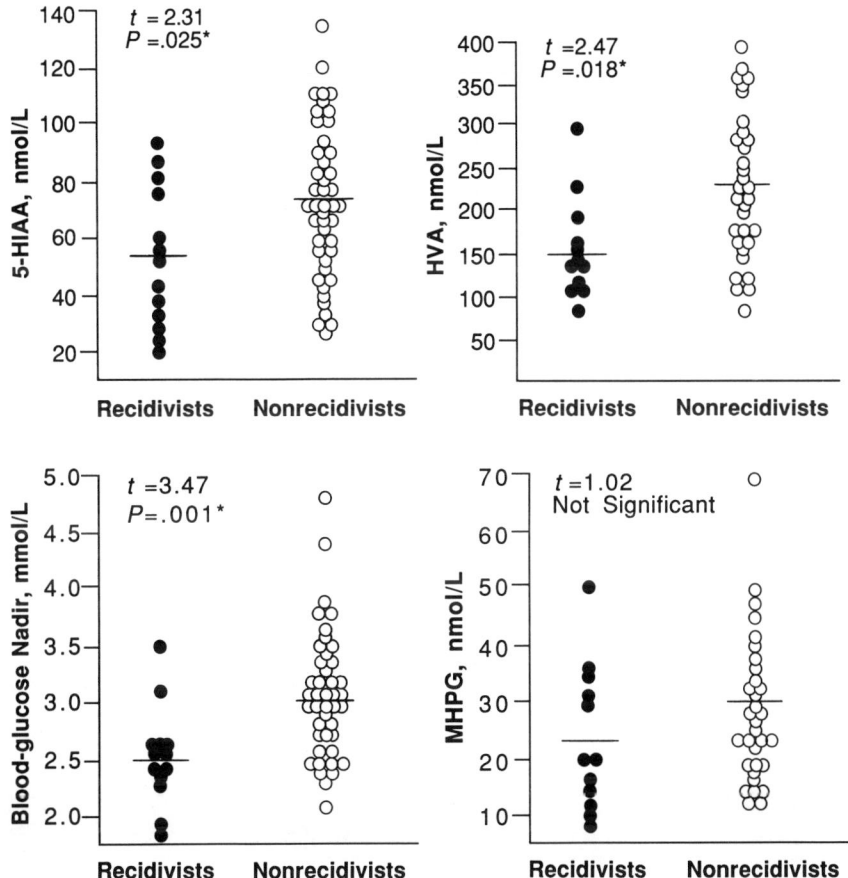

Figure 6.4. Cerebrospinal fluid monoamine metabolite concentrations and blood-glucose nadir in recidivist ($n=13$) and nonrecidivist ($n=45$) violent offenders and impulsive fire setters.

Finally, the discriminant analysis needs to be replicated in an independent sample to test its robustness. Even if the data are representative of only a subgroup of violent offenders and arsonists, they raise questions about the current treatment of these individuals. In Finland, they are usually convicted for manslaughter or arson and on the average are sentenced to seven years in prison.[28] According to the police reports and medical records, the individuals in the present sample committed their offenses almost without exception while under the influence of alcohol. They behaved well in prison, where alcohol is generally not available, and were released early because of good behavior. When released from prison, many of them started to drink excessively again, only to commit another serious offense.

The present results suggest that the individuals most likely to commit repeat offenses are identifiable with a relatively high likelihood using behavioral and psycho-biological variables. Due to their generally low CSF 5-HIAA concentrations, they may be treatable with serotonergic drugs. Such treatment could ameliorate a serotonergic deficit and has been reported to reduce alcohol consumption in humans[29,30] and to improve abnormal glucose metabolism in experimental animals.[31]

REFERENCES

1. Asberg M, Traskman L, Thoren P: 5-HIAA in the cerebrospinal fluid: a suicide predictor? *Arch Gen Psychiatry* 1976;33:1193–1197.
2. Agren H: Symptom patterns in unipolar and bipolar depression correlating with monoamine metabolites in the cerebrospinal fluid. II. Suicide. *Psychiatry Res* 1980;3:225–236.
3. van Praag HM: Depression, suicide and the metabolism of serotonin in the brain. *J Affective Disord* 1982;4:275–290.
4. Palanappian V, Ramachandran V, Somasundaram O: Suicidal ideation and biogenic amines in depression. *Indian J Psychiatry* 1983;25:286–292.
5. Banki CN, Arato M, Papp Z, Kurcz M: Biochemical markers in suicidal patients: investigations with cerebrospinal fluid amine metabolites and neuroendocrine tests. *J Affective Disord* 1984;6:341–350.
6. van Praag HM: CSF 5-HIAA and suicide in non-depressed schizophrenics. *Lancet* 1983;2:977–978.
7. Ninan PT, van Kammen OP, Scheinin M, Linnoila M, Bunney WE Jr, Goodwin FK: CSF 5-hydroxyindoleacetic acid in suicidal schizophrenic patients. *Am J Psychiatry* 1984;141:566–569.
8. Brown GL, Ebert NH, Goyer PF, Jimerson DC, Klein WJ, Bunney WE, Goodwin FK: Aggression, suicide, and serotonin relationships to CSF amine metabolites. *Am J Psychiatry* 1982;139:741–746.
9. Agren H: Life at risk: markers of suicidality in depression. *Psychiatr Dev* 1983;1:87–104.
10. Roy-Byrne P, Post RM, Rubinow DR, Linnoila M, Savard R, Davis D: CSF 5-HIAA and personal and family history of suicide in affectively ill patients: a negative study. *Psychiatry Res* 1983;10:263–274.
11. Berrettini WH, Nurnberger JI Jr, Linnoila M, Narrow W, Scheinin N, Seppala T, Simmons-Alling S, Gershon ES: CSF and plasma monoamines and their metabolites in euthymic bipolar patients. *Biol Psychiatry* 1985;20:257–260.
12. Brown GL, Goodwin FK, Ballenger JC, Goyer PF, Najor LF: Aggression in humans correlates with cerebrospinal fluid amine metabolites. *Psychiatry Res* 1979;1 :131–139.
13. Linnoila M, Virkkunen M, Scheinin M, Nuutila A, Rimon R, Goodwin FK: Low cerebrospinal fluid 5-hydroxyindoleacetic acid concentration differentiates impulsive from nonimpulsive violent behavior. *Life Sciences* 1983;33:2609–2614.
14. Virkkunen M, Nuutila A, Goodwin FK, Linnoila M: Cerebrospinal fluid monoamine metabolites in male arsonists. *Arch Gen Psychiatry* 1987;44:241–247.
15. Virkkunen M: Reactive hypoglycemic tendency among habitually violent offenders: a further study by means of the glucose tolerance test. *Neuropsychobiology* 1982;8:35–40.
16. Virkkunen M: Insulin secretion during the glucose tolerance test among habitually violent and impulsive offenders. *Aggressive Behav* 1986;12:303–310.
17. Asberg N, Nordstrom P, Traskman-Benz L: Biological factors in suicide. In Roy A, ed. *Suicide*. Baltimore: Williams and Wilkins; 1986:47–71.
18. Roy A, Agren H, Pickar D, Linnoila M, Doran AR, Cutler MR, Paul SN: Reduced CSF concentrations of homovanillic acid and homovanillic acid to 5-hydroxyindoleacetic acid ratios in depressed patients: relationship to suicidal behavior and dexamethasone nonsuppression. *Am J Psychiatry* 1986;143:1539–1545.

19. Agren H, Mefford IN, Rudorfer NV, Linnoila M, Potter WZ: Interacting neurotransmitter systems: a non-experimental approach to the 5-HIAA–HVA correlation in human CSF. *J Psychiatr Res* 1986;20:175–193.

20. American Psychiatric Association Committee on Nomenclature and Statistics: *Diagnostic and Statistical Manual of Mental Disorders*. 3rd ed. Washington, DC: American Psychiatric Association; 1980.

21. Scheinin M, Chang W-H, Kirk K, Linnoila M: Simultaneous determination of 3-methoxy-4-hydroxyphenylglycol, 5-hydroxyindoleacetic acid and homovanillic acid in cerebrospinal fluid with high-performance liquid chromatography using electrochemical detection. *Anal Biochem* 1983;131:246–253.

22. *SPSSx Users Guide*. New York: McGraw-Hill; 1983.

23. Monahan J: The prediction of violent behavior: toward a second generation of theory and policy. *Am J Psychiatry* 1984;141:10–15.

24. Notandon C, Harding T: The reliability of dangerousness assessments: a decision-making exercise. *Br J Psychiatry* 1984;144:140–155.

25. Weller NPI: Aspects of violence. *Lancet* 1987;2:615–617.

26. Soothil KL, Pope PJ: Arson: A 20-year cohort study. *Med Sci Law* 1973;13:127–138.

27. Lewis NDC, Yarnell H: *Pathological Firesetting (Pyromania). Nervous and Mental Disease Monographs*. Vol 82. New York: The Coolidge Foundation; 1951.

28. Lakimies Liiton Kustannus: *Suomen Laki I*. Helsinki: Valtion Painatuskeskus; 1987.

29. Naranjo CA, Sellers EN, Roach CA, Woodley DV, Sanchez-Craig M, Sykora K: Zimelidine-induced variations in alcohol intake by nondepressed heavy drinkers. *Clin Pharmacol Ther* 1984;35:374–381.

30. Naranjo CA, Sellers EM, Sullivan JT, Woodley DV, Kadlec K, Sykora K: The serotonin-uptake inhibitor citalopram alternates ethanol intake. *Clin Pharmacol Ther* 1987;41:266–274.

31. Nutt DJ, Gleiter CH, Linnoila M: Repeated electroconvulsive shock normalizes blood glucose levels in genetically obese mice (CS7BL/6J ob/ob) but not in genetically diabetic mice (CS7BL/KsJ db/db). *Brain Res* 1988;448:377–380.

7

Carbohydrates and Depression
Serotonin and Seasonal Affective Disorder

Richard J. Wurtman and Judith J. Wurtman

Abstract: Several related behavioral disorders recognized in the past decade are characterized by disturbances of appetite and mood. One of the best known is seasonal affective disorder, or SAD. In this chapter, two leading researchers show the complex interrelationship between serotonin, carbohydrates, and melatonin in the explanation of this frequently observed condition.

On May 16, 1898, the intrepid Arctic explorer Frederick A. Cook made the following notation in his journal:

> The winter and the darkness have slowly but steadily settled over us. . . . It is not difficult to read on the faces of my companions their thoughts and their moody dispositions. . . . The curtain of blackness which has fallen over the outer world of icy desolation has also descended upon the inner world of our souls. Around the tables . . . men are sitting about sad and dejected, lost in dreams of melancholy from which, now and then, one arouses with an empty attempt at enthusiasm. For brief moments some try to break the spell by jokes, told perhaps for the fiftieth time. Others grind out a cheerful philosophy; but all efforts to infuse bright hopes fail.

We now know that the members of the Cook expedition were suffering from classic symptoms of winter depression, a condition related to a recently described psychiatric disease known as seasonal affective disorder, or SAD. As the journal entry makes clear, recognition of the association between depression and the onset of winter is not new. But in recent years there has been growing interest in SAD and in two behavioral disorders, carbohydrate-craving obesity (CCO) and premenstrual syndrome (PMS), that share some of its symptoms. The symptoms include depression, lethargy and an inability to concentrate, combined with episodic bouts of overeating and excessive weight gain; they tend to be cyclic, recurring at characteristic times of the day (usually late afternoon or evening in CCO), month (just prior to menstruation in PMS) or year (generally fall and winter in SAD).

Over the past decade a wealth of information has emerged that casts light not only on the clinical expressions of this group of mood and appetite disorders but also on the disturbed biochemical processes that underlie them. It now appears that these disorders are affected by biochemical disturbances in two distinct biological systems. One system involves the hormone melatonin, which affects mood and subjective energy levels; the other involves the neurotransmitter serotonin, which regulates a person's appetite for carbohydrate-rich foods. Both systems are influenced by photoperiodism, the earth's daily dark-light cycle. Indeed, photoperiodism appears to be the basis for the cyclic patterns of all three disorders.

At high latitudes in the Northern and Southern hemispheres SAD appears in the late fall or early winter and lasts until the following spring. Once expressed, it tends to recur annually unless the patient moves to a place where day length does not decrease significantly in fall and winter. Sufferers complain of episodic bouts of depression combined with profound cravings for carbohydrate-rich foods. They go to sleep early and stay in bed for nine or 10 hours, unlike patients with nonseasonal depression, who have difficulty sleeping. Their sleep, however, is intermittent and not fully refreshing; during the day they are often drowsy and have trouble concentrating. Once spring arrives SAD patients are full of energy and creativity; they are almost manic in their zest for life. At the same time their craving for carbohydrates lessens and most lose the weight they had gained over the winter (figure 7.1).

The following case history typifies many SAD sufferers. Patient M, a 53-year-old teacher, stands five feet four inches tall and weighs 181 pounds. She is unhappy about her weight and over the years has spent a lot of money on short-lived diets. "I know my problem is carbohydrates: when I'm on a diet I stay away from bread, potatoes and sweets and I always lose weight. But when I'm not dieting I get anxious and tense in the mid afternoon and I'm unable to concentrate on what I'm doing. I want to eat something to calm myself, so I buy crackers or donuts and nibble on them. At home sometimes I just keep eating until I go to bed." Shortly after Thanksgiving, Patient M experienced two months of feeling tired and depressed. "I told my husband to leave me alone and assigned my pupils problem sets so I wouldn't have to talk to them at school. The house was a mess. I stopped eating except for bread and pasta, but I still gained weight. Finally when spring came I felt better perhaps because the school year was ending and summer was about to begin."

The symptoms described by Patient M are virtually the same as those associated with CCO and PMS, except that carbohydrate cravers are affected daily, typically in the late afternoon and early evening, and PMS sufferers are affected monthly, during the luteal phase of the ovarian cycle, which lasts for two weeks prior to the onset of each menstrual period.

Interest in seasonal mood disorders was sparked in the early 1980's, when Peter S. Mueller, a psychiatrist at the National Institute of Mental Health, reviewed data on

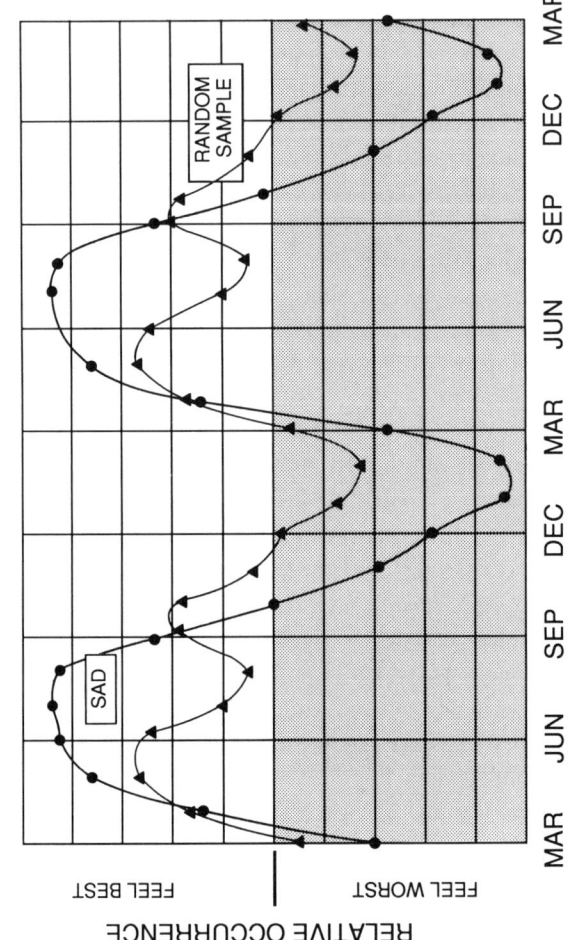

Figure 7.1. Seasonal fluctuations in mood in New York City.
Seasonal fluctuations in mood are common among people in New York City (and other northern areas) but are severest in patients diagnosed with SAD. The data are from a study by Michael Terman of the New York State Psychiatric Institute.

a 29-year-old woman he had been treating for cyclic bouts of winter depression. Over the course of several years the patient moved to a number of different cities. Mueller maintained contact with her and observed that the farther north she lived, the earlier she became depressed in the fall and the longer she stayed depressed in the spring. On two occasions when the woman traveled to Jamaica in midwinter her depression disappeared within a couple of days of arrival.

Mueller began to speculate that sunlight (or the lack of it) contributed in some way to the woman's depression and decided to experiment with phototherapy (a form of treatment previously shown to be effective in treating jaundiced infants and psoriasis). On consecutive mornings he exposed the patient to 2,500 lux of supplemental, full-spectrum light. (A lux is a unit equivalent to the illumination cast on a surface by one candle one meter away, which is equal to from one-fifth to one-tenth of a foot-candle.) In less than a week the patient had recovered from her depression.

Mueller's findings came to the attention of Norman E. Rosenthal, Thomas A. Wehr and Alfred J. Lewy, also at the NIMH, who were interested in the various manifestations of clinical depression. They launched a full-scale investigation into the natural history of winter depression, recruiting large numbers of volunteers for observation and treatment. The results were both revealing and intriguing. They confirmed the therapeutic effect of supplemental light in treating winter depression with phototherapy. In addition their data provided the first link between winter depression and carbohydrate craving.

A subsequent study by Steven G. Potkin, Daniel F. Kripke, William Bunney and their colleagues at the University of California at Irvine provided more complete data on the correlation in the U.S. between SAD and latitude. A questionnaire published in the newspaper USA Today in March of 1985 provided a description of SAD but omitted any reference to its presumed association with day length. Readers were asked to provide yes or no responses to 15 statements thought to characterize the disease. Those who responded yes to eight or more statements (and thus presumptively had SAD) were asked to send the questionnaire to the authors; 723 did so. The prevalence of SAD in each state was determined by dividing the number of respondents by average daily sales of the newspaper in that state. Results indicated that 100 people per 100,000 in the northern regions of the U.S. are affected by SAD; in the south the incidence is less than six people per 100,000 (figure 7.2). These estimates, however, are undoubtedly low because people with SAD are less likely to read newspapers and to answer questionnaires than unaffected people.

At about the same time, we began to investigate eating disorders at the Massachusetts Institute of Technology's Clinical Research Center, an inpatient clinic on the university campus. A typical study at the CRC might last for two weeks and focus on carbohydrate consumption among 20 patients in one of two weight groups: moderately obese (from 20 to 39 percent above ideal body weight) and obese (those who are from 40 to 80 percent above ideal body weight).

The eating habits of our study subjects were closely monitored both at regularly scheduled meals and between meals. Snack intake was measured by a computer-operated vending machine (based on a design by J. Trevor Silverstone of St. Bartholomew's Hospital Medical College in London) that was available around the clock and

PREVALENCE OF SAD IN THE UNITED STATES

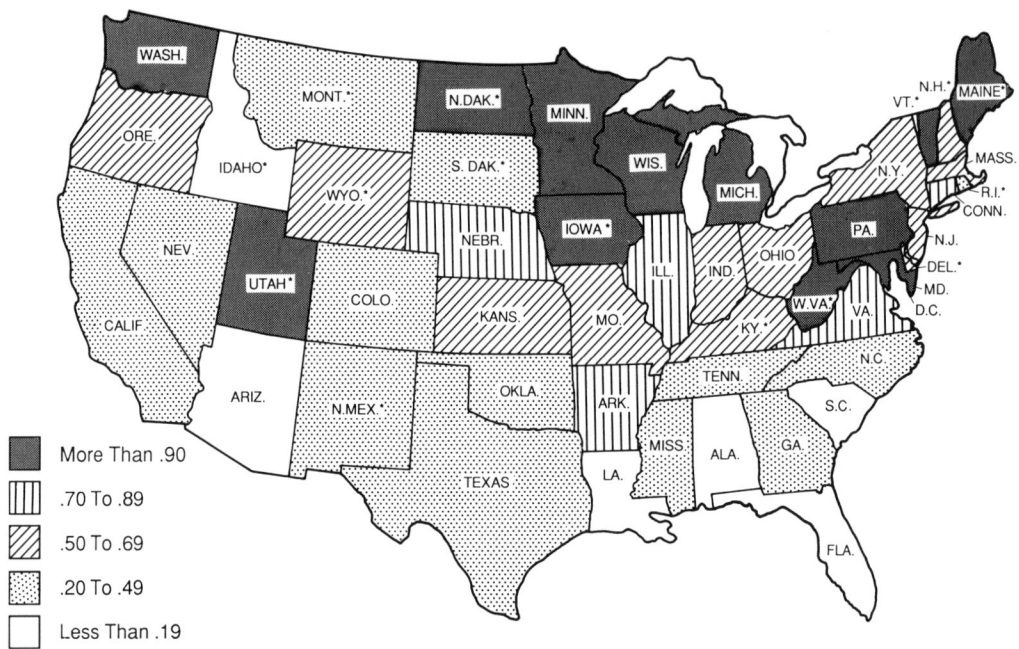

Figure 7.2. Prevalence of SAD in the United States.
Prevalence of SAD in the U.S. varies with latitude. In a northern state such as Minnesota, SAD affects more than 100 people per 1,000, whereas in Florida it affects few than six people per 100,000. Asterisks indicate a sample that is too small to be reliable. The date were collected by Steven G. Potkin and his associates at the University of California at Irvine.

contained a variety of snack foods ranging from carbohydrate-rich cookies to protein-rich sardines. All the selections contained roughly equal amounts of fat (six grams, for example) and calories (about 110). The foods could be obtained only by typing a special access number into a keyboard connected to a computer that kept a continuous record of the number and type of snacks selected by each patient. Participants in the study were asked to eat as they normally would and not be embarrassed about their calorie intake; most cooperated, believing the data we obtained would eventually help them to overcome their weight problem.

Food consumption during regular meals was measured by giving participants unlimited portions of food in preweighed, labeled containers that were color-coded and set on a table in the dining room. The different foods, like the snacks in the vending machine, varied in their protein and carbohydrate content and were equal to one another in fat and calories. At the end of each meal a dietitian reweighed the containers (and their leftovers) to determine how much of each food type a person had eaten.

Our studies at the CRC have enabled us to test and discard a number of myths concerning obesity, specifically with regard to carbohydrate craving. Most prominent among them, perhaps, is the notion that all obese people overeat anything that is tasty, whenever it is available. Instead it appears that those who are carbohydrate cravers overeat only carbohydrates and do so only at characteristic times of the day. At mealtime they behave like normal eaters, consuming a total of some 1,940 calories per day. (An average adult female consumes a total of from 1,500 to 2,000 calories, a male from 2,200 to 2,700 calories.) Toward the late afternoon or early evening, however, the volunteers begin to snack, often consuming an additional 800 or more calories per person per day. A similar pattern has been observed among women with PMS: they increased their snack intake by about 460 more calories per day than women at the same stage of the menstrual cycle who are not affected by PMS.

We were also intrigued to discover (with the help of the computer-run vending machine) that patients almost invariably underreport their consumption of snacks. It seems that a snack, if eaten quickly, is easily forgotten, as if it somehow "doesn't count." Yet for those concerned about their weight, snacks do count. In some cases they provide 30 percent or more of an individual's caloric intake.

Moreover, we found that most of the snacks consumed by CCO and PMS patients are carbohydrates (figure 7.3). We observed, in fact, that more than half of the

PROPORTION OF CALORIES AND CARBOHYDRATES CONSUMED AS SNACKS BY SAD PATIENTS

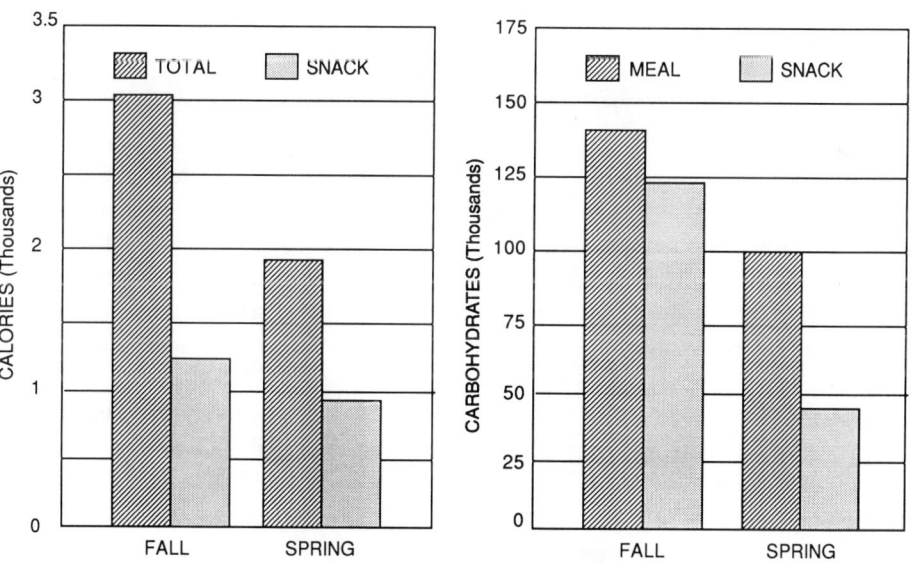

Figure 7.3. Proportion of calories and carbohydrates consumed as snacks by SAD patients. Proportion of calories and carbohydrates consumed in the form of snacks of SAD patients varies enormously depending on the season. In the fall patients consume more than 3,000 calories per day, of which about 1,200 are from snacks; in the spring their caloric intake falls below 2,000, of which fewer than 500 come from snacks (*left*). A similar pattern is apparent for carbohydrate consumption. In the fall almost 50 percent of the carbohydrates eaten per day are in the form of snacks; in the spring the proportion drops to roughly 30 percent (*right*).

carbohydrate-craving obese people at the CRC never select a protein snack although most will readily eat proteins at mealtime. A possible (but still unproved) explanation for such selective eating behavior is that in carbohydrate cravers the ability to regulate nutrient intake is impaired in the late afternoon or early evening. In a noncraver the desire for something sweet is infrequent, noncyclic and readily appeased, say by eating a cookie or two; in a carbohydrate craver, however, the desire may continue unabated until nine or 10 cookies are eaten. This suggests there is a malfunction in the feedback mechanism by which the brain knows carbohydrates have been eaten. Another possibility is that carbohydrate cravers snack not because they are hungry but because carbohydrate-rich foods improve their mood. Why snacking takes place at certain times of day for CCO patients is not clear; its cyclic occurrence, which is monthly in PMS or seasonal in SAD, may reflect the actions of ovarian hormones or melatonin on the brain, but no such relationship has been established for CCO. It is clear, in any case, that carbohydrate snacks tend to exacerbate obesity because they are often rich in fat and thus in calories. It appears that carbohydrate craving is a multifaceted disorder. As many as two-thirds of all obese people are carbohydrate cravers, but not all carbohydrate cravers are obese; many control their weight by exercising, eating low-calorie meals or satisfying their craving with such low-fat carbohydrates as popcorn (without butter) or candies such as jelly beans. Conversely, not all obesity is linked to carbohydrate craving. Some obese people show no preference for carbohydrates, and some overeat chiefly at mealtime, consuming snacks infrequently.

Our research also focused on mood fluctuations among carbohydrate cravers. When these people were given standardized psychiatric tests based either on an interview (the Hamilton Scale) or a written questionnaire (the Beck Depression Inventory), a high susceptibility to clinical depression was revealed. When carbohydrate cravers were asked why they succumb to foods they know will exacerbate their obesity, their explanation sounded much like the one provided by SAD sufferers. It almost never had to do with hunger or with the taste of the food; instead most said they eat to combat tension, anxiety or mental fatigue. After eating, the majority reported feeling calm and clearheaded. We wondered whether the consumption of excessive amounts of snack carbohydrates leading to severe obesity might not represent a kind of substance abuse, in which the decision to consume carbohydrates for their calming and antidepressant effects is carried to an extreme at substantial cost to the abuser's health and appearance. With the help of Harris R. Lieberman and Beverly R. Chew of M.I.T., one of us (Judith Wurtman) set out to test the relation between carbohydrate snacking and mood. Forty-six volunteers, including both carbohydrate cravers and noncravers, were given standard psychological tests before and after eating a carbohydrate-rich, protein-free meal. The carbohydrate cravers were significantly less depressed after snacking, whereas noncravers experienced fatigue and sleepiness. These findings suggest that carbohydrate cravers may eat snacks high in carbohydrates in order to restore flagging vitality, much as some people will pour another cup of coffee when they feel that their energy level or attention span is flagging.

The discovery that one's carbohydrate craving, like SAD, has a distinct periodicity led us to believe photoperiod might in some way be linked to the cyclic manifestations of appetite and mood disorders. We knew from work carried out some 25 years ago that the secretion of melatonin follows a distinct circadian rhythm coupled to daily and seasonal changes in light, which seems to match, at least conceptually, the rhythm most associated with SAD. Melatonin was discovered in 1958 by Aaron B. Lerner and his colleagues at the Yale University School of Medicine, who isolated it from the pineal glands of cattle and found that it lightened excised pieces of tadpole skin. Five years later Julius Axelrod and one of us (Richard Wurtman), then at the NIMH, suggested that melatonin was a hormone in mammals, based on its ability to suppress gonadal function when injected into rats. Subsequently we found that melatonin synthesis decreased when rats were exposed to light and that this effect was mediated by interactions among the retina, the brain and special sympathetic nerves that innervate the pineal gland [see "The Pineal Gland," by Richard J. Wurtman and Julius Axelrod; SCIENTIFIC AMERICAN, July, 1965].

At about the same time, Wilbur B. Quay of the University of California at Berkeley demonstrated that melatonin levels in the pineal gland of rats exhibit a daily rhythm, rising at night and falling during the day. A few years later Russell Pelham and his colleagues at the University of Pittsburgh described similar fluctuations of melatonin in the plasma of humans. Soon thereafter one of us (Richard Wurtman) and Harry J. Lynch of M.I.T. found that melatonin levels in human urine exhibit pronounced time-dependent fluctuations in samples taken from the same subjects: they are at least five times higher at night than they are during the day.

In order to prove that the timing of melatonin rhythms in humans is affected by the day-night, light-dark cycle, David C. Jimerson of the NIMH, Lynch and one of us (Richard Wurtman) examined the effects of abruptly shifting a person's photoperiod. We recruited a number of volunteers, monitored their plasma and urinary melatonin rhythms and then changed their photoperiod. We kept them indoors and on the test day left the lights on until 11:00 A.M., shifting the daily dark period 12 hours—to between 11:00 A.M. and 7:00 P.M.

We found it took four or five days for the subjects to reentrain and adjust physiologically to the new light cycle by secreting melatonin when it was dark and suppressing its secretion when it was light. Thus we showed that melatonin secretion follows a circadian rhythm in humans, as it does in other mammals, that the rhythm is endogenous (generated by a clock somewhere in the brain) and that it is entrained by the light-dark cycle (figure 7.4).

Neither we nor other investigators were able to demonstrate in humans, however, what Axelrod and one of us (Richard Wurtman) had shown more than a decade earlier in rats: that melatonin secretion is acutely suppressed if subjects are exposed to light during the dark part of the cycle. Perplexed, we concluded that the pineal gland of humans was inexplicably insensitive to the effects of light.

It was not until 1980 that Lewy discovered that melatonin secretion in humans can be acutely suppressed by light if the light is of sufficient intensity. When the participants in his study were awakened at 2:00 A.M. and exposed to 2,500 lux for one and a half hours, their plasma-melatonin levels declined abruptly. Thus light has two effects on melatonin rhythms in humans, just as it does in rats. It can either reentrain the melatonin rhythm (as when daytime was artificially reversed in our experimental

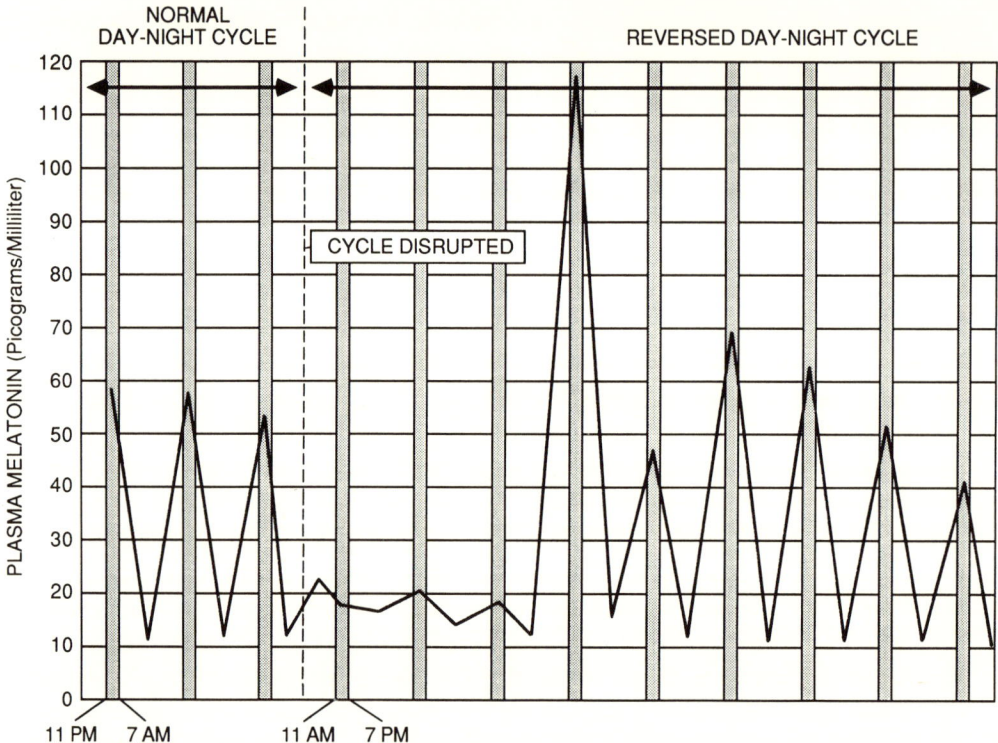

Figure 7.4. Daily rhythm of melatonin secretion.
Melatonin secretion follows a daily rhythm in humans, as it does in other mammals. During the day (*white columns*) secretion of the hormone from the pineal gland is suppressed and levels of melatonin in the plasma are low. At night (*dark columns*) melatonin is released from the pineal gland and its levels in the plasma are high. If the daily light cycle is abruptly shifted by 12 hours so that the dark period runs from 11:00 A.M. to 7 P.M. (instead of 11:00 P.M. to 7:00 A.M.), several days are needed before the new cycle of melatonin secreation reentrains (adjusts) to the new photoperiod. Once adjusted, melatonin secretion again follows a rhythmic pattern.

study) or suppress melatonin secretion entirely (if the dark period is eliminated). Either action or both could underlie light's therapeutic effect on SAD.

The work of Mueller, Rosenthal and others demonstrated that exposing SAD patients to intense supplemental light for a few hours each morning could eliminate their depression and carbohydrate craving after a few days. Obese carbohydrate cravers have not yet been treated with phototherapy, but a preliminary study by Barbara L. Parry of the NIMH suggests that supplemental light may be effective in treating women with PMS, whose symptoms worsen in the winter.

Michael Terman of Columbia University has found that exposing SAD patients to 2,500 lux for two hours in the morning brings complete remission from both depression and carbohydrate craving in roughly half of them, usually after only a few days of treatment. Most of the remaining patients show some improvement short of full remission. Although his study is not yet complete, Terman thinks it may be possible to enhance the efficacy of treatment by increasing the amount of time patients are exposed to light or by increasing the intensity of the light, say to 10,000 lux. Certainly such levels would more closely approximate sunlight, which ranges from 10,000 lux on a cloudy day in northern Europe to 80,000 lux on a sunny day close to the Equator. Other investigators, however, propose that it is the duration of phototherapy, rather than its timing, that is important in treating SAD. In any case it is now clear that the light must be at least 2,500 lux; customary indoor lights (which range in intensity from 250 to 500 lux) suppress neither the symptoms of SAD nor melatonin synthesis.

Researchers find that light administered in the morning is more effective than light administered later in the day. That finding has been interpreted by Lewy, Terman and others as an indication that light advances a person's circadian rhythm and shortens the dark phase of melatonin secretion. Terman and his associates have noted that the decline in plasma melatonin, which normally occurs early in the morning, is delayed in SAD patients by about two hours. Perhaps high-intensity light induces clinical remission when it is administered in the morning by shortening the daily period of melatonin secretion by several hours.

Is SAD caused by melatonin—either too much of it or when it is secreted for too long? Or is melatonin simply a convenient indicator for another process that underlies the disease? At the moment we cannot answer that question, but circumstantial evidence does suggest a direct link between melatonin and SAD. Lieberman, Lynch and one of us (Richard Wurtman) found that the administration of rather large doses of melatonin to normal individuals induces sleepiness, decreases alertness and slows reaction time. Perhaps the onset of melatonin secretion in the evening is an important promoter of sleep, sensitizing the brain to other sleep-inducing factors. That may explain why SAD patients are hypersomnic in winter, when the daily dark period is almost twice as long as it is in spring. A link between melatonin and mood is also suggested by the ability of oral melatonin to worsen a patient's depression; unfortunately no drug has been developed that selectively blocks melatonin's production or its actions.

But why do patients with SAD, CCO and PMS have a tendency to crave carbohydrate snacks? Why is it that only some people are vulnerable to CCO? And how is it that the brain normally knows when carbohydrates have been or should be consumed? Inhabitants of developed countries habitually eat from 12 to 14 percent of their calories in the form of protein and about three or four times as much in the form of carbohydrates. Even a bear will eventually forsake honey for an occasional fish. How is such a phenomenon regulated? We now know that the answer to these questions involves serotonin, one of the neurotransmitters: substances that are released from a neuron when it fires and that convey the nerve impulse across the synapse to the next neuron.

Serotonin is a derivative of tryptophan, an amino acid that is normally present at low levels in the bloodstream. The rate of conversion is affected by the proportion of carbohydrates in a person's diet: carbohydrates stimulate the secretion of insulin, which facilitates the uptake of most amino acids into peripheral tissues, such as muscle. Blood tryptophan levels, however, are unaffected by insulin and so the proportion of tryptophan in the blood relative to the other amino acids increases when carbohydrates are consumed. Since tryptophan competes with other amino acids for transport across the blood-brain barrier, insulin secretion speeds its entry into the central nervous system, where it enters, among other cells, a special cluster of neurons known as the raphe nuclei. There it is converted into serotonin (figure 7.5).

The level of serotonin in turn figures in a feedback mechanism affecting the amount of carbohydrate an individual subsequently chooses to eat [see "Nutrients That Modify Brain Function," by Richard J. Wurtman; SCIENTIFIC AMERICAN, April, 1982].

SEROTONIN REGULATION OF CARBOHYDRATE CONSUMPTION

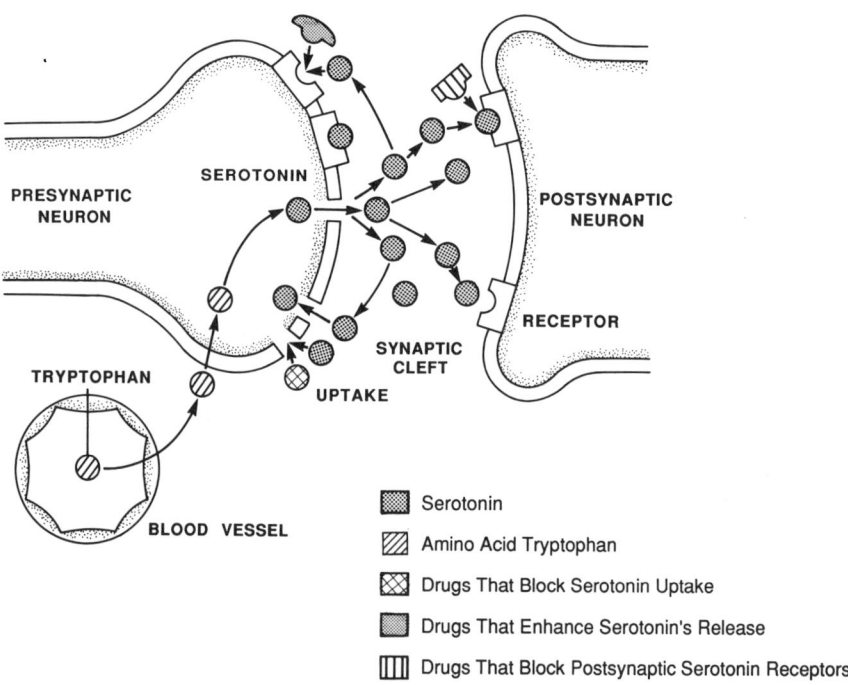

Figure 7.5. Serotonin regulation of carbohydrate consumption.
Serotonin regulates carbohydrate consumption. The process begins with the amino acid tryptophan (▨), which circulates through the blood to the brain, where it enters the raphe nuclei. After entering a presynaptic neuron, tryptophan is converted by way of a two-step process into serotonin (▦). Serotonin is then released into the synaptic cleft separating the presynaptic neuron from the postsynaptic neuron. Serotonin that reaches the postsynaptic neuron binds to special receptors. Serotonin levels rise in response to carbohydrate consumption. As more serotonin is released, more information is thus transferred to the postsynaptic neuron, where it activates a feedback mechanism. When its concentration is high, serotonin binds to presynaptic receptors, thereby suppressing the release of additional serotonin from the presynaptic neuron. It can also be rapidly removed from the synapse by uptake into the presynaptic neuron. Drugs that enhance serotonin's release (▦) or that block its reuptake (▧) increase information transfer across the synapse and diminish carbohydrate snacking; drugs that block postsynaptic serotonin receptors (▥) increase appetite, particularly for carboyhydrates.

When the feedback mechanism is disturbed, as we believe happens cyclically in patients with SAD, CCO and PMS, the brain fails to respond when carbohydrates are eaten, and so the desire for them persists longer than it should.

Serotonin regulates other behaviors too, including mood and sleepiness. Bonnie Spring, now at the University of Health Sciences/Chicago Medical School, found that noncarbohydrate-craving women become sleepy and prone to committing errors following the consumption of a high-carbohydrate lunch (which is expected to increase brain serotonin levels). Similar responses among noncarbohydrate-craving obese individuals were noted by Lieberman and one of us (Judith Wurtman). In contrast, carbohydrate cravers reported feeling refreshed and invigorated after eating a meal rich in carbohydrates.

The mechanisms affecting the relative proportions of carbohydrate and protein in one's diet are most apparent when feedback loops are disrupted, as they are when a patient is given drugs that affect serotonin-mediated neurotransmission. Rats that are allowed to choose between two or more synthetic foods containing different proportions of carbohydrate and protein will normally alternate between them. If, however, the rats are either injected directly with serotonin (into the brain) or given drugs that enhance the effect of serotonin by promoting its release into nerve synapses, prolonging its activity or stimulating its receptors, then carbohydrate intake in experimental rats is selectively reduced.

Drug trials carried out on humans show that a serotoninlike drug, d-fenfluramine (which releases serotonin into brain synapses and then prolongs its action by blocking its reuptake into the presynaptic neuron), has a similar effect, selectively suppressing carbohydrate snacking in patients affected by CCO. We also found, while collaborating with Dermot A. O'Rourke, a psychiatrist at the Massachusetts General Hospital, that d-fenfluramine can also be effective in treating SAD: it reduces carbohydrate snacking (and its associated weight gain) while simultaneously ameliorating the symptoms of depression. More recently, with Amnon Brzezinski of the Hebrew University–Hadassah Medical School in Jerusalem, we found that d-fenfluramine may also be effective in treating similar symptoms in patients with PMS. In 12 of 17 individuals studied, administration of the drug over a six-month period led to a reduction in both carbohydrate craving and depression. Another disorder, which we think may be linked to serotonin (and thus to SAD, CCO and PMS), is a form of bulimia that is associated with severe binging, often on carbohydrate-rich foods, but with little or no vomiting. Most such patients are mildly obese women; many are severely depressed and come from families with histories of depression and alcohol abuse. Preliminary studies by G. F. M. Russell of the University of London and Arthur G. A. Blouin of the University of Ottawa suggest that d-fenfluramine can be effective in treating such women; those that respond to the antidepressant effects of the drug are most likely to benefit from its effects on appetite suppression. In general we have found that drugs that selectively facilitate serotonin-mediated neurotransmission (such as d-fenfluramine, femoxetine, fluoxetine, zymelidine and fluvoxamine) tend to cause weight loss, whereas drugs that block serotonin-mediated transmission or antidepressants that interact with neurotransmitters other than serotonin have the opposite effect: they often induce carbohydrate craving and subsequent weight gain.

No one could reasonably claim that the symptoms of SAD, CCO or PMS are inconsequential. Prolonged periods of deep depression and irritability can sorely compromise a person's ability to sustain essential human relations. But surely it is not abnormal to feel one's spirits flagging in the fall, to sometimes crave chocolate or pasta, to put on a few pounds every winter or to feel grumpy when beset by menstrual cramps. Indeed, seasonal changes in behavior afflict normal people as well as those with SAD. Among 200 subjects chosen at random from the New York City telephone directory and surveyed by Terman and his associates, half said they are less energetic in the fall and winter. Forty-seven percent said they gained weight during those months, 31 percent said they slept more and 31 percent said they were not as interested in social activities. Among respondents who reported a decline in energy at some time during the year, about 50 percent said their slump occurs in the fall and winter; only 12 percent said it occurs in summer. Terman concluded that a significant percentage of New York's population suffers from a mild form of SAD; we suspect that the inhabitants of other northern cities, such as Boston or Minneapolis, are similarly affected.

In Tromsø, Norway, which at a latitude of 69 degrees does not see the sun rise above the horizon between November 20 and January 20, midwinter insomnia is thought to affect 24 percent of the population. Charles S. Mullin, Jr., of the U.S. Naval Academy has described widespread sleeplessness, depression, irritability, impaired cognition and the gain of as much as from 20 to 30 pounds among scientists and military personnel who overwinter in Antarctica. The fact that SAD reaches its peak incidence in the Southern Hemisphere in June and July, incidentally, indicates that it is not simply a form of holiday blues or the result of melancholy reflections occasioned by the ending of another year.

Is SAD, then, merely an exaggeration of the normal human response to diminishing light levels in fall and winter? Is it perhaps analogous to hibernation? Probably not. Hibernating animals characteristically lower their body temperature, cease reproductive activity and spend the winter in deep sleep. People with SAD do none of those things; if anything, the time they spend in deep sleep (measured by electroencephalogram) is reduced. Perhaps contemporary lifestyles increase vulnerability to seasonal depression by diminishing the amount of time we expose ourselves to light: Daniel Kripke and his fellow workers measured the amount of time per day that healthy elderly subjects in San Diego—a region of particularly favorable climate—were exposed to sunlight. Surprisingly, the men were in sunlight for only 75 minutes out of each 24-hour period, the women for only 20 minutes. We need not all live in California, but perhaps most of us need to be exposed to more light, as our ancestors were. Perhaps much as office workers join health clubs to compensate for the lack of exercise, people with indoor jobs need to arrange for adequate exposure to light.

Much remains to be learned about mood and appetite disorders and about the link between serotonin and melatonin. Why does a SAD patient, for example, respond equally well to supplemental lighting, which presumably acts by affecting melatonin, and to drugs that stimulate the release of serotonin? And where might those treatments act in the sequence of pathophysiologic processes leading to SAD? Before we can answer those two questions, it would help to know whether or not light or melatonin has a direct effect on serotonin-releasing neurons. Until we have better answers, we can at least be grateful for the fact that these disorders respond to novel and effective therapies—even if the mechanisms by which the therapies work remain a mystery.

FURTHER READING

Seasonal Affective Disorder: A Description of the Syndrome and Preliminary Findings with Light Therapy. Norman E. Rosenthal, David A. Sack, Christian Gillin, Alfred J. Lewy, Frederick K. Goodwin, Yolande Davenport, Peter S. Mueller, David A. Newsome, and Thomas A. Wehr in *Archives of General Psychiatry*, Vol. 41, No. 1, pages 72–80; January, 1984.

D-Fenfluramine Selectively Suppresses Carbohydrate Snacking by Obese Subjects. Judith Wurtman, Richard Wurtman, Sharon Mark, Rita Tsay, William Gilbert and John Growdon in *The International Journal of Eating Disorders*, Vol. 4, No. 1, pages 89–99; February, 1985.

On the Question of Mechanism in Phototherapy for Seasonal Affective Disorder: Considerations of Clinical Efficacy and Epidemiology. Michael Terman in *Journal of Biological Rhythms*, Vol. 3, No. 2, pages 155–172; 1988.

Part 3

Factors Influencing Serotonin Function

*The complex interactions among individual experience and develop-
ment, genetics, and the role of serotonin contradict the nature-
nurture dichotomy that dominates most public discussions of neuroscience and behavor.*

8

Ontogeny, Social Experience, and Serotonergic Functioning

Benson E. Ginsburg

Abstract: Social experience plays an essential role in the expression of genes associated with serotonin and its interaction with other neurotransmitters. Contrary to the established dichotomy that has long dominated debates about human nature, it is neither "nature" nor "nurture" that is primarily responsible for much of our social behavior: as in other mammals, it is the interaction between an organism's inherited capacities and life history that establish the diverse responses making one individual so different from another.

Our understanding of behavior involves description, analysis, and synthesis at various levels, each of which has become a specialty unto itself. What is acceptable in human behavior is culturally dependent, but culture, as it affects behavior, may also be viewed as arising from "human nature." Human ethology, in the narrow sense of the term, is comparative, historical, and descriptive. Its comparative aspects include considerations of its evolutionary roots, arising from ethological and sociobiological studies of animal societies; as well as anthropological descriptions of primitive human societies; sociological studies of more advanced cultures (up to and including the present); and historical studies of records of human events—all involving descriptions and varying levels of analyses of human group behavior.

The human social world is partitioned by geography, natural and man-made ecologies, separation of gene pools, differences in sex roles, xenophobic tendencies, and variations in behavioral norms that are, in part, referable to these conditions. Are there, however, biological universals that interact with these variable factors and, to some degree, shape them? In simpler terms, is there a human nature that manifests itself statistically in spite of the wide occurrences of individual differences, which may also be substantially rooted in biology? Are our varying norms of social behavior based on interactions between our biological tendencies and the varying conditions under which we exist? Or do our social control mechanisms, as codified in religious and secular law, arise out of a need to regulate the expression of our biological tendencies while at the same time representing a human expression of social hierarchy that differs primarily in complexity from what we find in social organizations in the animal world?

In the span of one generation, we have seen a revolution in human behavioral norms that, while culturally based, reflect how universal and persisting tendencies have come to alternative expressions. In this country, the sexual revolution, civil rights, the feminist movement, and the spread of fundamentalist religion, as well as the shift in ethnic composition, have brought about momentous changes. Each of these topics is a dissertation in itself, but their common thrust has been to change behavioral norms by means of coalitions of like-minded groups with common interests. These are behaviors with ethological roots that have proven effective in animal societies as well.

Such changes have also impacted our medical and legal systems. Homosexuality, for example, was formerly considered to be an illness, meriting research effort directed toward the goal of effective treatment. It is no longer considered an illness. The legal status of abortion is again in flux due to organized social pressures, including the rights of the father in addition to the rights of the woman who is bearing the child. Alcoholism has been shown to have a genetic etiology, as have a number of psychopathologies and other handicapping conditions, such as Down's syndrome, autism, some forms of mental retardation, and other human behavioral variations (Vandenberg, Singer, & Pauls, 1986). The dysphoric aspects of premenstrual syndrome, where twin studies are suggestive of a genetic vulnerability, represent another case in point (Dalton, 1987). This has only recently been regarded as a possible psychiatric illness (see American Psychiatric Association, 1987), thereby potentially changing its status under the law.

To what extent does our legal system codify custom and help to maintain a hierarchical social structure? How are legal codes and sanctions changed, and how in particular can a knowledge of the evolutionary roots of human social behavior and the biosocial determinants of individual behavior help us to arrive at an optimal system of social responsibility? Should those who conceive of themselves as outside the system (as did those in the Tawana Brawley case) be free to ignore its procedures and sanctions, and are they, in fact, the victims of an imposed social hierarchy enforced by laws designed to maintain it? In this chapter, I deal with possible homologies in animal societies; the relationship between social status and aggressive behaviors; how biology affects individual and group behavior; and how social relationships throughout life interact with biological predispositions to evoke or change behavioral outcomes. Since it is aggressive and violent behavior that is of urgent social and legal concern, a knowledge of the mechanisms associated with such behaviors becomes particularly pertinent. At this level, we are here most concerned with several of the key neural transmitters associated with agonistic behaviors and especially with serotonin. This concern should not be seen as simple reductionism but rather as an attempt to focus on one important biological mechanism, as well as to place it in these broader contexts.

ETHOLOGICAL ROOTS OF SOCIAL BEHAVIOR

Science is hydra-headed. On the one hand, it represents the most certain knowledge that we have, while on the other hand, it is constantly changing. However, when it comes to studies of the evolution of social behavior, we remain in the Darwinian tradition. Following this tradition, the normative social behavior of a species is viewed as genetically based and as the product of natural selection, conferring adaptive advantages to both the individual and the group. Ethologists, in particular, have focused on social behavior and social organization and found it to be hierarchical. The cement that holds it together consists of social bonding and social dominance (Buck & Ginsburg, 1991). Disrupting the ontogeny of normal social experience in many animal groups is seen as interfering with the ability of the affected individuals to develop species-typical behavior. It has also been shown that these latter biological tendencies, or "set points," remain latent and that compensatory experiences can often restore their expression (MacDonald & Ginsburg, 1981). We see the human counterpart of the animal studies on disruptive experience in such manifestations as the lasting effects of child abuse, as one example, and the capacity to return to normality under appropriate compensatory regimes, or even spontaneously. Where compensatory regimes must be instituted, this is covered by the rubric of "therapy." The identification of the developmentally disruptive factors is a product of scientific research, as is the development of effective modes of therapy.

Abnormal developmental circumstances are not the only causes of later aberrant behavior. Genetic variations as well are often involved, usually interactively, and occasionally, at least with our current state of knowledge, as irrevocable determinants. Genetically hyperaggressive individuals have been found in most species where this aspect of behavior has been studied. These include mice, rats, guinea pigs, rabbits, dogs, and domestic fowl, among others (Ginsburg, 1967; Ginsburg & Carter, 1987). In each, selective breeding has been shown to affect dramatically agonistic behavior, often within only a few generations of selection. In most species studied, selection experiments increase the levels of aggression, thereby implying, if not demonstrating, that natural selection has not favored the extremes of aggression that are potentially available in the gene pool of a species (Ginsburg & Carter, 1987). In fact, when such behavior arises, the hyperaggressive individual does not become dominant in the social hierarchy and is usually eliminated from the group. This has been dramatically shown in studies of wolf packs, where, in the case of males, an overdominant individual is generally deposed by a coalition, and in the case of females, by a sometimes-lethal attack on the part of a previous subordinate (Jenks & Ginsburg, 1987). This would argue, therefore, that ethological studies do not support the proposition that we are, as a species, violent by nature. While assertive and aggressive tendencies may be innate, violence is taught (Ginsburg & Carter, 1987).

We have also amassed considerable evidence demonstrating that differences in aggressive behavior, while genetically based, depend on developmental circumstances for their degree and mode of expression (Scott and Fredericson, 1951). Early-handling paradigms in mice and rats constitute such variables. Isolation enhances aggression in male mice (Ginsburg, 1966; Valzelli, 1969). Separation from littermates in aggressive pedigrees within some dog breeds prevents the expression of that tendency, while in dingoes, similar manipulations do not produce the same effect (Fisher, 1955; Ginsburg & Carter, 1987). It has been shown that the sensitive period during postnatal, prepubertal development for altering aggressive tendencies is itself genetically variable, as is the direction and magnitude of the behavioral change during the optimal period for inducing it. In mice, for example, some genotypes will show augmented aggression under the same regime that will reduce aggression in others and may have no effect on still other genetic substrata (Ginsburg, 1968b). These examples provide tools for investigating the mechanisms leading to heightened aggression, their variations within species, and the possibilities of predicting and affecting outcomes.

Xenophobic tendencies and examples of organized group aggression have been dramatically described in chimpanzees, particularly in the Gombe, where the displacement and annihilation of one group by another has been graphically documented (Goodall, 1986). Equally graphic is the description of the restoration of normal behavior in chimpanzees that have been transplanted to a normal ecological environment after having passed varying portions of their lives, including early infancy through adulthood, in behaviorally aberrant circumstances (Carter, 1988). While the attainment of the species-typical norm or set point took time and "therapy" to achieve, this was nevertheless accomplished. Instructive nonhuman ethological examples are, therefore, available to us and suggest that we, too, have species-typical behaviors, from which there are genetic and environmentally induced variations and for which there are both sanctions and therapies. The overdominant wolf, for example, becomes an isolate or is killed. However, these and other forms of socially disruptive behaviors can be studied in species ranging from mice to man where, perhaps not surprisingly, at the neural and genetic levels, we find similar mechanisms.

Science is constantly changing, and the newer knowledge of neural mechanisms associated with aggressive behavior is one example of the advances in our understanding of biological bases of this class of behaviors.

SCIENCE AND THE LAW

Before turning to the details of this newer knowledge in a particular subspecialty, it is well to remind ourselves that the scientific enterprise has a dual relationship with the law. On the one hand, "scientific" findings have often been overhastily applied—as in the case of various eugenic programs, including castration for violent offenders, most notably in sexual crimes, or in the use of prefrontal lobotomies to control aggression. Controversies remain regarding the use of electroconvulsive therapies, and many additional examples could be adduced. Where scientific knowledge has shown promise for alleviating a condition leading to uncontrolled violence, as in some cases of late luteal-phase dysphoric disorder, the medical model for rehabilitation, even were it to be well established, is often in conflict with the demands of society sanctioned by law, not only for compensation but for vengeance. As Jeffery (1987) has pointed out, our criminal justice system not only does little for rehabilitation but tends to perpetuate those behaviors that it should be designed to change or prevent, beginning with the juvenile offender, perhaps even earlier.

There is yet another aspect of science and the law in which the social responsibility of science is at issue. How far is it permissible to go when human testing is involved, as in the case of new drugs with potentially dangerous side effects? What about the use of fetal brain tissue in research for the treatment for Alzheimer's disease? Where are the boundaries of genetic engineering? What does behavioral science tell us about the parameters that should be considered, perhaps in legal codification, for surrogate mothering? The list can be geometrically expanded, even if we confine ourselves only to issues dealing with human behavioral problems.

On the positive side, advances in the biobehavioral sciences offer our best hope for creating social systems that extend to the home, the schools, and the medical and the legal professions and that optimize the social potentials within our societies, including therapeutic and corrective measures. What, then, is the relationship between our newer biobehavioral knowledge, extending from the population to the molecular level, and the broader issues that we are attempting to address?

BIOLOGICAL FACTORS IN AGGRESSIVE BEHAVIOR

In an evolutionary sense, aggression serves many useful purposes. Aggression does not necessarily involve fang-and-claw encounters. Territorial singing in birds, territorial markings in various mammals, and the establishment of nest sites all involve aggressive displays and also, subsequently, additional behaviors in relation to courtship, mating, predation, and many other functions, including social position and social role in hierarchical group organizations. Aggression also involves defense of the nest, defense of the young, and predatory behavior. Social signals often serve as substitutes for overt fighting. Stereotyped threat behaviors, warning calls, submissive behaviors, and many others that have been well studied and described have all, in one way or another, been considered part of the aggressive repertoire. Where communicative behaviors avoid actual encounters, they serve an obvious adaptive value to the individual, as well as to the group (Buck & Ginsburg, 1991). The hierarchical organization of many groups, however, may deprive lower-ranking individuals of a role in reproductive behavior while at the same time ensuring that, as a member of an organized group, those individuals have access to food and are otherwise sustained in a more optimal manner than if there were no social organization serving group functions. Such selectivity of reproductive roles serves to partition the gene pool within the population and to enhance the process of evolution by increasing genetic diversity upon which selective forces then act (Ginsburg, 1968a; Wright, 1939). Therefore, agonistic behaviors that include various categories (all of which may, in one way or another, be deemed aggressive) are triggered by different stimuli, involve different neural pathways, and are associated with differences in neurochemical mediators, which are, in turn, often themselves associated with particular neural structures and innervations.

In addition, hormonal factors are known to play a role both developmentally and proximally. In the human, the male hormone testosterone serves an organizing role in the brain prenatally, as well as later during puberty. On the molecular level, labeled steroid hormones can be followed and seen to enter the nuclei of cells within the nervous system and are known to activate gene expression in a selective manner. This can be dramatically and unequivocally shown in interfertile species hybrids, such as dog and coyote, where the defensive-threat behavior of the coyote is not within the genetic repertoire of the dog. In a hybrid possessing both genetic capacities, crosses involving genetic segregation have produced phenotypes that, from an early age and in the absence of learning models, exhibit the behavior characteristic of one species or the other. In animals carrying the genetic capacity for the coyote defensive-threat behavior along with the genetic system of the dog, the phenotypic segregants include animals that spontaneously switch from the dog behavior to the species-typical coyote behavior during or after puberty. The coyote genes have, in this instance, been brought to active expression by the sex steroids in both sexes, and the behavior itself is triggered when the animal is aggressively challenged.

In some segregants, where the coyote genetic system is known to be present, the pubertal hormonal changes are necessary but not sufficient to switch the repertoire from one genetic system to another. In such instances, subjecting the animals to continuing social stress by placing them in a living situation with more dominant individuals elevates the cortisol levels of the subordinates and, after a period of time and under challenge that must be aggressively met, the species-typical coyote defensive threat is manifested. Once the genes for this system have been turned on, they persist throughout the lifetime of the animal. In a preliminary experiment, mimicking the cortisol profile produced by social stress and subsequent aggressive challenge produces the same result in the appropriate genotype (Moon & Ginsburg, 1985; Moon, 1993). Here, environmental factors acting through hormonal systems selectively regulate gene expression. The behavior itself, however, is mediated through neural pathways that communicate via chemical messengers; and particular brain regions, as well as neural pathways, are associated with these component behaviors. Unlike the findings for squirrel monkeys, where the cortisol levels have been reported higher in dominant than in subordinate males, it is the subordinate animals under continual social stress who show the elevated cortisol levels. These may be situational differences rather than species differences. In mice, there are strain (i.e., genetic) differences in the level of brain enzymatic activity in the major substrate involved in serotonin production (Diez, Sze, & Ginsburg, 1976). These differences are associated with intermale aggression, as tested in paired encounters between previously isolated males. Further work with this neural transmitter has produced varying results. Some of these results may reflect species differences; others may reflect ecological or procedural differences; but in general, concentrating on a single neural mechanism where complex behaviors are involved is probably too simplistic. More recent studies have, therefore, considered the effect of serotonin situationally and in conjunction with the effect of other neural transmitters.

The work of McGuire and Raleigh (1987) with vervet monkeys adds still other dimensions to our consideration of the relationships between a particular neural transmitter and situational parameters. With respect to serotonin in whole blood and in the brain, they report a positive association with dominant status in males, particularly in the presence of subordinate males. They also show an increase with a change in social status when a male is rising in the dominance hierarchy. Pharmacologically induced decreases in serotonin appear to increase irritability and therefore intolerance to social interaction regardless of rank, resulting in an increase in aggressive behaviors. The serotonergic mechanism, therefore, is affected by and may help to mediate the social structure of the group, at least insofar as male vervet monkeys are concerned (Raleigh et al., 1991). The authors also cite studies that would lead us to expect similar results in squirrel monkeys and, inferentially, in humans.

Additional studies focusing on serotonin levels in other species have also reported associations between serotonin levels, serotonin turnover, and elements of aggressive behavior (McGuire & Raleigh, 1987; Miczek, Mos, & Olivier, 1989). While these results are not entirely clear-cut, there does appear to be an association between increased serotonin and decreased aggression, as well as the reverse. In some cases, the relationship is situation dependent, and certainly other mechanisms and transmitters are also involved. Because the blood levels change slowly, work with mice and rats has emphasized the changes occurring in the brain. While serotonin is simply one element involved in the mechanisms of the activation of aggressive behaviors, it does nevertheless have a demonstrable effect. Two additional lines of evidence support this: the first is that significant serotonergic innervation and changes in serotonin level associated with aggression are found in areas of the brain associated with such behaviors (Fink & Reis, 1981); the second is that the pharmacological manipulation of serotonin levels, binding, and turnover also modify the expression of such behaviors (Winslow & Miczek, 1983). Despite positive findings of an association between serotonin levels, serotonin turnover, and modifications of aggressive behavior, an association that is sometimes situational, there is also strong negative evidence that would dispute or modify the meaning of these associations. For example, Olivier and associates (Olivier et al., 1987) have reported that, in comparing an aggressive with a nonaggressive strain of rat, there was a lack of correlation in individual rats between the level of serotonin turnover in the striatum and several measures of aggressive behavior on which the strains differed. By contrast with other studies, their data suggest that there is a facilitative role for the neural transmitter in strain comparisons. However, there was no such demonstrable association with individual differences within strains.

In addition to the level and turnover of any neural transmitter in particular brain regions, there is the further question of where and how they are bound by specific receptors, which usually have characteristic distributions. In attempting to reason from the various levels and mechanisms of serotonin involvement in aggression, it is essential for the behavioral endpoints to be highly selective. Using ethological methodologies for this purpose, Olivier et al. (1987) developed a class of psychoactive drugs, which they term "serenics." These affect the serotonergic system primarily through the 5-hydroxytryptamine (5-HT) binding sites. They therefore conclude (p. 180) that "the general hypothesis that 5-HT activity is inversely correlated with aggression, seems untenable. A more refined [5-HT] hypothesis should specify a role for 5-HT receptors. . . . The presence of 5-HT receptors in several areas known to be involved in the modulation of agonistic behavior . . . lent support to a direct regulation of such behavior by 5-HT-related mechanism." Thus, 5-HT sites were present in high density in brain areas associated with aggressive behavior but with other behaviors as well. Other 5-HT receptors have been mapped to different distributions. Further work by the same investigators, using maternal aggression in rats as an endpoint, lends support to the role of the 5-HT receptor sites in the mediation of this behavior. Specific agonists to these receptors were the most effective in the reduction of aggression. These results are similar to those obtained with a resident-intruder paradigm in male rats. This work has been extended to effects of mouse-killing behavior in rats, play fighting in juvenile rats, and social interactions in forms as diverse as pigs and vervet monkeys. The effects appear to be primary rather than secondary because they occur in the absence of sedation or other sensory-motor disturbances. In addition, the drug-treated animals are capable of normal social interactions. Defense and flight capabilities are unaffected. Therefore, the investigators conclude that the mechanism of action of serotonin predominantly involves the 5-HT binding site and that this effect is involved specifically in the modulation of offensive aggression.

EFFECTS OF DOPAMINE

As mentioned earlier and acknowledged by all investigators in this field, studies for aggressive behavior deal with a complex set of phenomena. "Aggression" is not a unitary behavior. There are, for example, differences in the effects of pharmacological agents on offensive and defensive aggression, and these can also be separated in experimental animals using genetic analyses (Blanchard et al., 1979; Adams, 1980; Brain, 1981). In addition, there are situational factors that affect outcome. There is also the involvement of hormonal systems and of other neural transmitters, not to mention the effects of opiate peptides, neuromorphologic differences in innervations, all of which must interact with each other, as well as with environmental and eliciting variables. Since the focus of the "Serotonin, Social Behavior and the Law" conference was on a particular neurotransmitter, and since we must acknowledge that it is simply one mediator as a neurochemical messenger, it will be instructive to focus on another such mediator and then to consider how to progress from the detailed studies of how each importantly involved chemical messenger produces its effects, to integrating these in terms of combined effects, as well as considering how they relate to other neuroendocrine, physiological, and psychological factors. The ultimate question for purposes of the present discussion is whether detailed knowledge of one possibly critical factor involved in the mediation of aggression can be used to achieve better understanding, prediction, and control. It is this aspect that finally articulates with the social, legal, and medical aspects of the problem.

My colleague Professor Ruth Guttman, of the Hebrew University of Jerusalem, and I recently reported some results we had obtained using a presumed dopamine agonist on two mouse strains differing in their aggressive behavior as measured by testing isolate-reared males in dyadic encounters (Ginsburg & Guttman, 1988). A total aggression score was obtained based on all aggressive acts occurring during a set period of observations over two days. In addition, various components of the aggressive behaviors were separately identified and scored. Those of particular interest for this study included overt attack, chase, wrestle, and tail rattle. When testing the two strains without drugs, the total aggression score for the C57BL/6 strain was dramatically higher than that of the BALB/c. On the component scores, attack and chase cohered in the C57 mice and together accounted for over 90% of the agonistic behavior while being involved in only 28% of agonistic acts among the BALBs. In this strain, over 70% of the agonistic behaviors were accounted for by wrestle and tail rattle (table 8.1). On treatment with dl-amphetamine, the total aggression score for the C57 strain was halved, while that of the BALBs was doubled (table 8.2). The effect of this pharmacological agent is, therefore, contingent on the genotype of the subject. Further, in examining the component scores, the major reduction in aggressive acts for the C57 animals occurred in the attack and chase categories, while the major increase in the BALBs could be attributed to wrestle (table 8.2). In reciprocal F_1 crosses of these two strains, there was no differential effect in the untreated hybrids. However, on 5 mg/kg of dl-amphetamine, the reciprocally derived hybrids closely resembled their male parents (see figure 8.1). Based on these findings as well as other data, we have attributed this, in large part, to a Y-chromosome effect. These results have been replicated and extended by Westenberger, Ginsburg, and Guttman (1992).

Amphetamine has been shown to have a variety of behavioral effects and acts not only as a dopamine agonist but also in other ways and on other systems. In this instance, however, we concluded that the actions were largely dopaminergic and referable to the effects of dl-amphetamine as a dopamine agonist based on the following information. Pharmacogenetic studies with amphetamine have been used with various strains of mice and their crosses to investigate the relationship between genetics, dopamine, and behavior. Low-dose amphetamines have been found to stimulate locomotor activity, whereas higher doses have been associated with stereotypic behavior. The evidence is that the locomotor activity is mediated via the nucleus accumbens, whereas stereotypic behaviors are related to the striatum. Tyrosine hydroxylase, the rate-limiting enzyme in the production of dopamine, has been reported to be higher in BALB mice than in C57s (Baker, Joh, & Reis, 1982). These differences are not based on differences in the enzymatic activity per se but rather on the number of dopamine-containing neurons. F_1s between BALB and C57 mice have been reported to be intermediate in their tyrosine hydroxylase activity and to show simple segregation in F_2, thereby suggesting a simple genetic predisposition. Further work using recombinant strains derived from a BALB–C57 cross shows three groups with respect to tyrosine hydroxylase activity in the substantia nigra, further suggesting that relatively few genes are involved. In the striatum (associated with stereotypic activity), crosses between BALB and C57 mice showed dominance of tyrosine hydroxylase activity in the F_1. For these reasons, cited in Ginsburg and Guttman (1988), and because the divergent effects on component behaviors follow the differences to be expected from the differences in distribution of dopaminergic neurons, we have chosen to attribute these effects of amphetamine to its role as a dopamine agonist (Westenberger et al., 1992; Westenberger, 1993).

Comparing these results with those on serotonin illustrates that, if one looks at

Table 8.1. Patterns of Agonistic Behavior in Inbred and Crossbred Saline-Treated Control Mice

Strain	N	Total Score	Attack		Chase		Wrestle		Tail Rattle	
			Frequency	Percentage of Total	Frequency	Percentage of Total	Frequency	Percentage of Total	Frequency	Percentage of Total
C57BL/6	19	21.53	13.26	62	6.37	30	1.58	7	0.32	1
BALB/c	20	15.75	3.55	23	0.85	5	5.10	32	6.25	40
♀ C57 × BALB ♂	12	32.17	14.83	45	11.33	35	1.42	4	4.58	14
♀ BALB × ♂ C57	10	32.70	18.60	57	8.20	25	1.70	5	4.20	13
Combined F_1	22	32.41	16.55	51	9.90	31	1.55	5	4.41	14

Source: B.E. Ginsburg & R. Guttman, 1988. Genetic assortment of dopaminergic determinants of patterns of agonistic behavior in male mice. Paper presented at the 18th annual meeting of the Behavior Genetics Association, Nijmegen, The Netherlands, June.

Table 8.2. Patterns of Agonistic Behavior in Inbred and Crossbred Mice Treated with 5.0 mg/kg dl-Amphetamine

Strain	N	Total Score	Attack		Chase		Wrestle		Tail Rattle	
			Frequency	Percentage of Total	Frequency	Percentage of Total	Frequency	Percentage of Total	Frequency	Percentage of Total
C57BL/6	19	10.63	5.05	48	2.32	22	0.58	5	2.68	25
BALB/c	20	30.50	6.65	21	0.70	2	11.40	37	11.75	39
♀ C57 × BALB ♂	20	36.40	14.60	40	9.70	27	0.40	1	11.70	36
♀ BALB × ♂ C57	10	11.80	3.30	28	2.50	21	2.40	20	3.60	31
Combined F_1	20	24.10	8.95	37	6.10	25	1.40	5	7.65	32

Source: B.E. Ginsburg & R. Guttman, 1988. Genetic assortment of dopaminergic determinants of patterns of agonistic behavior in male mice. Paper presented at the 18th annual meeting of the Behavior Genetics Association, Nijmegen, The Netherlands, June.

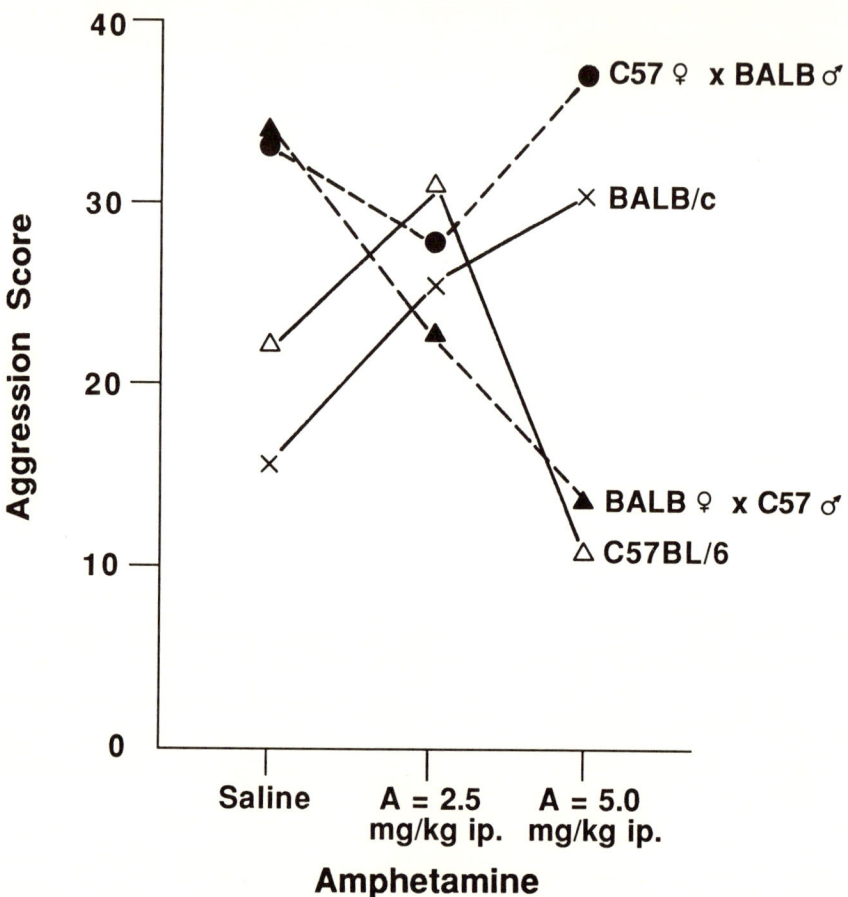

Figure 8.1. Mean total aggression scores. *Source:* B.E. Ginsburg & R. Guttman, 1988. Genetic assortment of dopaminergic determinants of patterns of agonistic behavior in male mice. Paper presented at the 18th annual meeting of the Behavior Genetics Association, Nijmegen, The Netherlands, June.

either system alone, it can be viewed as a major factor in the initiation and expression of aggressive behavior. An important dimension added by the work just cited is that, in addition to the other variables influencing the relationship between neural transmitters and their effects on behavioral endpoints, differences in the underlying genotype are also major factors. Where the effect on one genotype is to increase aggression, the effect on another is to decrease it. Were we then to examine a genetically variable population of mice, we could probably make a statistically accurate statement regarding the effect of a dopamine agonist on aggressive behavior. As a quantitative statement, it would merely be a reflection of which genotypes predominate in that population. With respect to the individual in the absence of any other information, the knowledge of the population norm becomes simply a statement of odds that cannot serve as a predictor for any specific instance. This is true not only for the dopaminergic system but for many other systems as well. An entire science of pharmacogenetics has grown around these differential responses to identical pharmacological treatments on seemingly identical behavioral endpoints.

Using this information as the basis for a hypothetical model, let us assume that pathological variations in human aggressive behavior can be referred to such mechanisms. On such a simplistic and reductionistic hypothesis, it could be argued that for some individuals treatment with the so-called serenic drugs would serve as palliatives and might shift the behavior to normal limits. In this case, at the level of mechanism, we are dealing with effects on binding sites for serotonin. In other cases, this mechanism could be entirely irrelevant, and tendencies toward socially unacceptable aggressive behavior could be conceived of as being mediated by another system—perhaps variations in dopaminergic innervation. It is also conceivable that these and still other systems act in concert and that there is an intervening variable involving mood or affect, which could, in turn, be referable to the opiate peptides. Clearly, the more we know about these systems, the better equipped we are to understand mechanisms at the neural level that are involved in the mediation of aggressive behavior and perhaps also in the differences in threshold that elicit its expression. Whether and how these systems are affected by cognitive mechanisms and by the priming effects of untoward early experience are not yet known. Based on the genetic data, we would expect that different individuals would be differentially susceptible to the effects of intervening events in their life histories. Based on the data in hand, the neuroendocrine mechanisms might be expected to be controlling links in the physiological chain of events mediating between experience, genotype, and neurophysiological and neurochemical final common paths.

CONCLUSIONS

On the basis of the complex interplay involving psychological, genetic, and biochemical variations in relation to aggressive behavior, how can we relate research at this level to the role of aggression in a human social context and to how we might best deal with this in law? The most informative and relevant example that comes to mind, perhaps because it is one with which I have dealt and continue to deal professionally, is that of the late luteal-phase dysphoric disorder affecting a small percentage of menstruating women (American Psychiatric Association, 1987). Here, we have an example of a biological variation in which some women cannot exercise normal control over aspects of their mood and behavior during the late luteal phase of the premenstruum. While for some women this may involve temporary alienation, depression, memory lapses, and other troublesome actions and feelings, for a few, there is uncontrolled hostility and anger, even leading to overt physical aggression (Dalton, 1987). In Britain, this biologically based aberration has been successfully used as an insanity defense, and now that late luteal-phase dysphoric disorder is on the threshhold of being recognized as a psychiatric entity in this country, we are likely to see some legal precedents established soon. The aspect of this disorder that is most relevant here is that science has so far been unable to isolate a particular cause or to identify a universally effective treatment. Given its association with the menstrual cycle, there is an obvious association with hormonal events. However, women who are troubled during the late luteal phase do not differ in their endocrine profiles from women who have no such difficulties. It has been postulated that it may be the response to these rapid hormone fluctuations involving genetic susceptibilities, perhaps through neural mechanisms including serotonin, that differentiate those women who are psychologically troubled at this time from those who are not (Ashby et al., 1988, 1990). Further, the nature of the types of biological variations that we have been discussing earlier may also determine whether the behavioral manifestations will be aggressive or otherwise. Hormonal treatments with such agents as progesterone have been effective in some women (Dalton, 1987), and Dalton cites some cases in Britain where the medical model has been successfully applied. Such women are effectively on parole, contingent on compliance with the treatment regime and with proof of its effectiveness. Twin studies and studies with first-degree relatives are suggestive of a genetic susceptibility for manifesting not only aberrant behavior but particular types of aberrant behavior for late luteal-phase dysphoric disorder. Such researches, while suggestive, are still fragmentary. There is also a growing literature indicating that women who experience severe distress at this time have been abused or otherwise traumatized as children (Paddison et al., 1990). Work in this area, therefore, suggests that there is a spectrum of biological vulnerabilities, that the triggering of the vulnerability may depend on priming by early experiential factors, and that, for some cases at least, effective treatments can be applied to bring the behavior under control.

Despite the complexities of the behaviors, the variations in biological vulnerability, and the psychosocial predisposing factors, a more detailed understanding of the underlying mechanisms involved in the mediation of aggressive behavior can lead us to diagnosis at the individual level and to effective intervention. For the serotonergic system, one can think of it as one of a number of possible pacemakers; and in those instances where it does represent a final common path or at least one that if altered would produce a better behavioral outcome, there is the possibility of treatment with serenic drugs. It is the responsibility of the scientific enterprise to provide us with a better understanding of these mechanisms. The baton can then be passed to other professions for using this knowledge to achieve ameliorations for conditions that are pathological for the individual and socially disruptive. The legal profession will then have the option of using this knowledge to justify reduced sentencing based on diminished capacity or even exoneration and remanding to treatment based on the insanity defense. Where ameliorative intervention can prove effective, the possibility of rehabilitation using a medical approach should be considered and incorporated as an integral part of our criminal justice system. As Jeffery (1987) has suggested, this will involve joint medicolegal efforts and changes, where appropriate, either in the legal code itself or in the way in which it is administered.

REFERENCES

Adams, D.B. (1980). Motivational systems of agonistic behavior in muroid rodents: A comparative review and neural model. *Aggressive Behavior, 6,* 295–346.

American Psychiatric Association. (1987). *Diagnostic and statistical manual of mental disorders* (3rd ed. rev.), pp. 367–369. Washington, DC: Author.

Ashby, C.R., Jr., Carr, L.A., Cook, C.L., Steptoe, M.M., & Franks, D.D. (1988). Alteration of platelet serotonergic mechanisms and monamine opidase activity in premenstrual syndrome. *Biological Psychiatry, 24,* 225–233.

Ashby, C.R., Jr., Carr, L.A., Cook, C.L., Steptoe, M.M., & Franks, D.D. (1990). Alteration of 5-HT uptake by plasma fractions in the premenstrual syndrome. *Journal of Neural Transmitters General Section, 79,* 41–50.

Baker, H., Joh, T.H., & Reis, D.J. (1982). Time of appearance during development of differences in nigrostriatal tyrosine hydroxylase activity in two inbred strains. *Developmental Brain Research, 4,* 157–165.

Blanchard, R.J., O'Donnell, V.O., & Blanchard, D.C. (1979). Attack and defense behaviors in the albino mouse. *Aggressive Behavior, 5,* 341–352.

Brain, P.F. (1981). Differential types of attack and defense in rodents. In P.F. Brain and D. Benton (Eds.), *Multidisciplinary approaches to aggression research* (pp. 55–77). The Netherlands: Elsevier Press.

Buck, R., & Ginsburg, B.E. (1991). Spontaneous communication and altruism: The communicative gene hypothesis. *Review of Personality and Social Psychology, 12,* 149–175.

Carter, J. (1988). Survival training for chimps. *Smithsonian, 19,* 36–51.

Dalton, K. (1987). Should premenstrual syndrome be a legal defense? In B.E. Ginsburg and B.F. Carter (Eds.), *Premenstrual syndrome: Legal and ethical implications in a biomedical perspective* (pp. 287–300). New York: Plenum Press.

Diez, J.A., Sze, P.Y., & Ginsburg, B.E. (1976). Tryptophan regulation of brain tryptophan hydroxylase. *Brain Research, 104,* 396–400.

Fink, J.S., & Reis, D.J. (1981). Genetic variations in midbrain dopamine cell number: Parallel with differences in responses to dopaminergic agonists and in naturalistic behaviors mediated by central dopaminergic systems. *Brain Research, 222,* 335–349.

Fisher, A.E. (1955). *The effect of differential early treatment on the social and exploratory behavior of puppies.* Unpublished doctoral dissertation, Pennsylvania State University, University Park.

Ginsburg, B.E. (1966). All mice are not created equal: Recent findings on genes and behavior. Teller Lecture, *The Social Service Review, 40,* 121–134.

Ginsburg, B.E. (1967). Social behavior and social hierarchy in the formation of personality profiles in animals. In J. Zubin and H.F. Hunt (Eds.), *Comparative psychopathology: Animal and human* (pp. 95–114). New York: Grune & Stratton.

Ginsburg, B.E. (1968a). Breeding structure and social behavior of mammals: A servo-mechanism for the avoidance of panmixia. In D.G. Glass (Ed.), *Genetics: Biology and behavior series* (pp. 117–128). New York: Rockefeller University Press and Russell Sage Foundation.

Ginsburg, B.E. (1968b). Genotypic factors in the ontogeny of behavior. *Science and Psychoanalysis, 12,* 12–17.

Ginsburg, B.E., & Carter, B.F. (1987). The behaviors and the genetics of aggression. In J.M. Ramirez, R.A. Hinde, and J. Groebel (Eds.), *Essays on violence* (pp. 59–76). Seville, Spain: Publicaciónes de la Universidad de Sevilla.

Ginsburg, B.E., & Guttman, R. (1988). Genetic assortment of dopaminergic determinants of patterns of agonistic behavior in male mice. *Behavior Genetics, 18,* 718 (abstract).

Goodall, J. (1986). *The chimpanzees of Gombe: Patterns of behavior.* Cambridge, MA: Belknap Press of Harvard University Press.

Jeffery, C.R. (1987). Criminal law, biological psychiatry, and premenstrual syndrome: Conflicting perspectives. In B.E. Ginsburg and B.F. Carter (Eds.), *Premenstrual syndrome: Legal and ethical implications in a biomedical perspective* (pp. 125–147). New York: Plenum Press.

Jenks, S.M., & Ginsburg, B.E. (1987). Socio-sexual dynamics in a captive wolf pack. In H. Frank (Ed.), *Man and wolf* (pp. 375–399). The Hague: Dr. W. Junk Publishers.

MacDonald, K.B., & Ginsburg, B.E. (1981). Induction of normal prepubertal behavior in wolves with restricted rearing. *Behavior and Neural Biology, 33,* 133–162.

McGuire, M.T., & Raleigh, M.J. (1987). Serotonin, social behaviour, and aggression in vervet monkeys. In B. Olivier, J. Mos, and P.F. Brain (Eds.), *Ethopharmacology of agonistic behaviour in animals and humans* (pp. 202–222). Dordrecht/Boston/Lancaster: Martinus Nijhoff Publishers.

Miczek, J.A., Mos, J., & Olivier, B. (1989). Brain 5-HT and inhibition of aggressive behavior in animals: 5-HIAA and receptor subtypes. *Psychopharmacology Bulletin, 25,* 399–403.

Moon, A. (1993). The ontogeny of expression of communicative genes in coyote x beagle hybrids. Unpublished doctoral dissertation, University of Connecticut, Storrs.

Moon, A., & Ginsburg, B.E. (1985). *Genetic factors in the selective expression of species-typical behavior of coyote × beagle hybrids.* Invited paper presented at the 19th International Ethological Conference, Toulouse, France.

Olivier, B., Mos, J., van der Heyden, J., Schipper, J., Tulp, M., Berkelmans, B., & Bevan, P. (1987). Serotonergic modulation of agonistic behaviour. In B. Olivier, J. Mos, and P.F. Brain (Eds.), *Ethopharmacology of agonistic behaviour in animals and humans* (pp. 162–186). Dordrecht/Boston/Lancaster: Martinus Nijhoff Publishers.

Paddison, P.L., Gise, L.H., Lebovits, A., Strain, J.J., Cirasole, D.M., & Levine, J.P. (1990). Sexual abuse and premenstrual syndrome: Comparison between a lower and higher socioeconomic group. *The Academy of Psychosomatic Medicine, 31,* 265–272.

Raleigh, M.J., McGuire, M.T., Brammer, G.L., Pollack, D.B., & Yuwiler, A. (1991). Serotonergic mechanisms promote dominance acquisition in adult male vervet monkeys. *Brain Research.*

Scott, J.P., & Fredericson, E. (1951). The causes of fighting in mice and rats. *Physiological Zoology, 26,* 273–309.

Valzelli, L. (1969). Aggressive behavior induced by isolation. In S. Garattinni and E.B. Sigg (Eds.), *Aggressive behavior* (pp. 70–76). New York: John Wiley and Sons.

Vandenberg, S.G., Singer, S.M., & Pauls, D.L. (1986). *The heredity of behavior disorders in adults and children.* New York: Plenum.

Westenberger, M. M. (1993). Neurogenetic basis of components of aggressive behavior in C57BL/6J and BALB/cBYJ mice and their reciprocal crosses. Unpublished doctoral dissertation, University of Connecticut, Storrs.

Westenberger, M.M., Ginsburg, B.E., & Guttman, R. (1992, July). A pharmacogenetic study of the effects of amphetamine on intermale aggression in mice. Poster presentation, 22nd annual meeting of the Behavior Genetics Association, Boulder, CO.

Winslow, J.T., & Miczek, K.A. (1983). Isolation of aggression in mice: Pharmacological evidence of catecholaminergic and serotonergic mediation. *Psychopharmacology, 81,* 266–291.

Wright, S. (1939). *Statistical genetics in relation to evolution.* Paris: Hermann.

9

Serotonin, Aggression, and Violence in Vervet Monkeys

Michael J. Raleigh
and
Michael T. McGuire

Abstract: A group's social behavior and an individual's dominance status play a central role in serotonin function: among primates, the dominant male has enhanced levels of serotonin that appear after status change as a consequence of perceiving others' submissive behaviors. For other group members as well, behavioral consequences, such as the initiation of aggressive behaviors directed at "inappropriate" objects, can depend on the way social experience produces changes in neurochemistry, which in turn modulate behavior.

We have long been interested in the bidirectional relationships between social behavior and brain monoaminergic function in vervet monkeys. One aspect of our research views social behavior as a cause of interindividual differences in physiological profiles. Among adult monkeys living in stable social groups, there are persistent differences in social status, affiliative interactions, and style of aggressive behavior. These individual differences in behavior induce many interanimal differences in biological processes. For instance, in several species of monkeys, high-ranking adult females are far more likely than other females to exhibit predictable, stable menstrual cycles that enhance the likelihood of becoming pregnant (1). Indeed in some species ovulation in subordinate females may be suppressed, perhaps due to stress-induced alterations in brain function that in turn lead to the absence of surges in luteinizing hormone (LH) (2).

Status-linked physiological differences are not confined to adult female primates. For example, in many monkey species, high- and low-ranking adult males differ in their basal and challenged testosterone concentrations. Similarly, dominant adult male vervet monkeys manifest whole-blood serotonin (WBS) levels that are approximately twice those of subordinate males. Many of these biochemical differences are state dependent, in that changes in social status (from subordinate to dominant or vice versa) are followed by corresponding physiological alterations (3). Among vervets and other monkeys, social factors not only influence hormonal titers but also affect brain serotonin concentration and function.

In another focus of our research, we examine the impact of biological factors on social and individual behavior. We have used two approaches to examine the behavioral consequences of alterations in neural function. One set of studies is experimental and involves pharmacological interventions. These investigations evaluate the effects of small or moderate alterations in serotonergic function on normal or species-typical patterns of behavior. Our intent has been to alter the rates and sequences of regularly occurring behaviors, such as grooming or aggression, by experimental manipulation of serotonin levels.

The other approach is correlational and involves measurement of peripheral biochemical parameters. These measures are selected because they provide information about brain chemistry and function without compromising the behavior of socially living animals. The results from both sets of studies are strikingly consistent and demonstrate that, in vervet monkeys, increased serotonin not only inhibits destructive aggression but also facilitates a wide range of positive, prosocial behaviors. Almost all of the conclusions about the impact of serotonin on social behavior derived from these studies can be generalized to humans. Consequently, vervets represent an appropriate animal model in which many aspects of the links between serotonin and behavior can be addressed more directly than they can be in humans. For instance, potentially critical variables, such as differential consumption of alcohol, can be more precisely controlled in monkeys than in humans. Studies of monkeys also facilitate long-term or prospective investigations, which assess the relationship between serotonin and behavior throughout an individual's lifetime.

In this chapter, we review aspects of the relationship between social behavior and serotonin in vervet monkeys. Because much attention has been focused on unlawful and destructive behavior, our discussion emphasizes those behaviors that might serve as analogues to human lawlessness. In particular, we concentrate on destructive aggression directed at other members of the same social group. Among vervets and other monkeys, destructive aggression occurs rarely but may have dramatic consequences, including ostracism, reproductive suppression, and the infliction of severe wounds (4). In monkeys, the probability of this type of aggression occurring is positively linked to social instability, alcohol consumption, and adolescence (5). In our view, this is the type of aggression that most closely parallels interpersonal violence among humans.

Our investigations, as well as observations by others, support the view that serotonin modulates and inhibits destructive aggression. Serotonergic systems seem to exert much of their inhibitory impact on aggression by diminishing the activity of other neuronal systems that facilitate aggression. Although robust, the inverse link between serotonin and aggression is not invariate; rather, it is affected by social factors.

In this chapter, the primary social variable we consider is an individual's social status or dominance rank. Clearly, a host of additional social factors influence the link between serotonin and violence. These include, but are far from limited to, the social status of the subject's mother, the subject's early rearing experiences, the animal's temperament, the situation eliciting the behavior, and the stability and predictability of the current social group. Thus, we regard our discussion of the relationships between serotonin and social status as illustrative of an approach to examining a complex, multiply caused relationship between behavior and physiological function. In no way do we view social status as the sole, or even the primary, environmental factor shaping the tie between aggression and neurobiology.

In the remainder of this chapter, we enumerate biological parameters that may serve as indexes of central serotonergic function, comment on the nature of dominant and subordinate relationships among nonhuman primates, present physiological data linking decreased serotonergic function to heightened aggression and violence, review pharmacological studies that implicate diminished serotonin as a cause of inappropriate aggression, and speculate on some of the ways animal studies may clarify aspects of human behavioral biology.

INDEXES OF SEROTONERGIC FUNCTION

Because of our interest in ongoing patterns of social behavior, the health and well-being of the monkeys we study is of central importance. This orientation imposes certain constraints on the procedures we can use to assess central serotonergic function. In our long-term studies, information about serotonergic function has been derived from three sources.

1. Hormones
   Cortisol
   Prolactin
   Growth hormone
2. Platelet properties
   Blood serotonin
   Receptor characteristics
   Serotonin uptake
   Monamine oxidase (MAO) activity
3. Cerebrospinal fluid 5-hydroxyindoleacetic acid (CSF 5-HIAA)

The concentrations of selected hormones may provide useful clues about central serotonergic function. In clinical studies, the titers of cortisol, growth hormone, and prolactin are sometimes used as indirect indexes of serotonergic function. The rationale underlying the use of these measures is that brain serotonergic activity influences the concentration of these hormones. An advantage of these measures is that they can be routinely collected and readily assayed. However, several drawbacks accompany their use. The concentration of any of these hormones is influenced by a multitude of factors. Prolactin, for instance, is affected by gender, age, sleep patterns, weight loss, stress, and (among females) lactational status. Further, while serotonin facilitates prolactin secretion, dopamine exerts a much stronger inhibitory effect (6). Thus, interindividual differences in prolactin may arise from many factors, and identification of the precise contribution of serotonergic systems to these differences is far from straightforward. Similar interpretative difficulties accompany the use of cortisol and growth hormone concentrations as indirect indexes of serotonergic function.

A second class of measures includes whole-blood serotonin (WBS) and other platelet properties. These measures are examined because they may mirror central serotonergic function. For example, blood serotonin undergoes the same biosynthetic pathway as does brain serotonin. Tryptophan is converted to 5-hydroxytryptophan (5-HTP), which in turn is decarboxylated to serotonin. Both blood and brain serotonin are sequestered in storage vesicles and undergo similar metabolism. Thus, because similar factors control aspects of their production, uptake, and storage, blood serotonin may serve as a marker of brain serotonin (7).

Because serotonin in the blood does not cross the blood-brain barrier, blood and brain serotonin are not necessarily related (see Yuwiler, Brammer, and Yuwiler, chapter 4, this volume). For instance, patients with carcinoid syndrome have exceedingly high levels of blood serotonin with no apparent change in brain serotonin concentration. A related concern is that brain serotonin seems to be less responsive to alterations in precursor (tryptophan) availability than is blood serotonin. Further, in the brain, tryptophan is converted into serotonin in serotonergic neurons. In contrast, blood serotonin is not synthesized in platelets. Rather, it is synthesized in the gut and taken up and stored in circulating platelets. Blood serotonin may therefore be more reflective of gut physiology than of brain function. Similarly, the life cycle of platelets is very different from that of neurons, and there are important metabolic and morphological contrasts between platelets and neurons. Thus, the link between brain serotonergic function and blood serotonin may be somewhat tenuous (8).

We have also used another peripheral measure as an index of serotonergic function: cerebrospinal fluid 5-hydroxyindoleacetic acid (CSF 5-HIAA). CSF 5-HIAA is the principal metabolite of brain serotonin and thus has substantially more validity as an index of brain serotonergic activity than the other two sets of measures (9). Nonetheless, there are several significant limitations involved in the interpretation of CSF 5-HIAA activity. Potential regional differences in brain serotonergic function cannot be addressed. In many species, serotonin is differentially distributed throughout the brain. In humans and vervets the concentration of serotonin is much higher in the limbic system than in the cerebral cortex. It is likely that serotonergic neurons in some areas (e.g., the amygdala) are more potently involved in mediating aggression and other emotionally charged behaviors than are those in other areas (e.g., the cerebellum) (10). However, at present it is not possible to identify which set of serotonergic neurons is primarily responsible for the concentration of CSF 5-HIAA. In addition to the absence of regional specificity, CSF 5-HIAA accumulates over time. Consequently, the temporal link between alterations in serotonergic activity and CSF 5-HIAA may be difficult to delineate. Acute changes in brain serotonergic function may or may not lead to parallel rapid changes in CSF 5-HIAA. Finally, at a practical level, CSF collection poses a greater risk to humans and monkeys than does blood collection.

Despite these and other drawbacks, CSF 5-HIAA has enormous utility as an index of brain serotonergic activity. In addition to its validity, CSF 5-HIAA concentration has two features that make it particularly well suited to our research. The first is that within individual adult subjects, CSF 5-HIAA concentrations are remarkably stable over time. Table 9.1 shows the mean and standard error of the concentration of CSF 5-HIAA from 12 adult males. Each male was sampled at least four times over a 10-week period. The table illustrates that there was very little within-subject variability. For no subject was the standard error of the mean greater than 14%.

A second feature of CSF 5-HIAA is that the differences between individuals are substantial. Among adult vervet monkeys there is about a threefold range in the concentration of CSF 5-HIAA, and this range is not due to the presence of one or two outliers. Thus, CSF 5-HIAA is a reliable biological marker, and it is appropriate to attempt to link individual differences in behavior or personality with CSF 5-HIAA concentrations.

Table 9.1. Intrasubject
Stability of CSF 5-HIAA

Subject	CSF 5-HIAA
1	52.8 ± 1.8
2	53.8 ± 5.0
3	56.0 ± 7.8
4	56.2 ± 2.2
5	62.0 ± 3.0
6	62.8 ± 2.6
7	71.9 ± 3.4
8	76.8 ± 2.4
9	77.5 ± 5.4
10	81.3 ± 4.8
11	85.2 ± 4.2
12	89.9 ± 5.3

Note: CSF samples were obtained from 12 adult male subjects living in stable social groups. Each animal was sampled at least four times over a 10-week period. CSF 5-HIAA concentrations (μg/mL) are in the mean ± standard error.

DOMINANT AND SUBORDINATE SOCIAL STATUS

Early studies of nonhuman primates tended to intermingle the concepts of aggression and dominance. Based largely on early accounts of terrestrial baboons, the popular literature promoted the notion that the largest, most aggressive individual was the highest ranking. As an increasing number of species living in a variety of habitats were examined, it became clear that dominance and aggression were not necessarily synonymous. Dominance currently refers to sustained asymmetrical relationships between two or more group members. Operationally, dominance relationships are recognized in two ways. One relies on approach-retreat interactions. If one member of a pair of animals (A) consistently displaces the other (B), then A would be regarded as dominant to B. The second criterion is success in, but not the rate of initiating, agonistic encounters. When one member of a pair routinely defeats the other, the reliable winner is regarded as the dominant member of the dyad. Among vervets and other species, results from these two procedures correlate highly, and there is rarely any difficulty in recognizing high- and low-ranking animals (11).

Table 9.2 highlights the relationships among dominance, morphological features, and aggression in adult vervet monkeys. The table summarizes observations made of socially living, captive vervet monkeys. Each of the characteristics was measured in at least 12 groups, and in each group the highest ranking (or most dominant) male or female was compared to a gender- and age-matched subordinate group member. For each feature, the dominant animal's score was divided by that of a matched subordinate. The table presents the mean percentage (and standard error) of these ratios. For example, on average, dominant males weighed 94% as much as subordinate males, and dominant females weighed 96% as much as subordinate females. As

Table 9.2. Morphological and Behavioral Concomitants of Dominance Status

Characteristic	Female	Male
Weight	96 ± 11	94 ± 12
Canine length	102 ± 9	109 ± 9
Testes volume	—	110 ± 12
Length	106 ± 14	107 ± 12
Aggressivity	84 ± 7*	74 ± 13*
Reconciliation	139 ± 11*	146 ± 10*
Recruit allies	203 ± 24*	199 ± 21*
Defend group	145 ± 19*	136 ± 12*

Note: This table shows the relationship between social status and selected morphological and behavioral measures in adult vervet monkeys. For each measure, the scores of dominant adult females and males are expressed as the percentage (± standard error) of age-matched subordinate animals. Each cell is based on data from at least 12 different social groups. An asterisk indicates $P <$ .05.

illustrated by the table, there are no striking morphological differences between dominant and subordinate adult animals. Neither dominant females nor dominant males weighed more, possessed larger canines, or had longer crown-rump measures than did subordinate animals.

While there were no striking status-linked morphological differences, the table does show that there were substantial rank-related behavioral contrasts. Relative to subordinate females, dominant females were less likely to initiate aggression, more likely to be involved in reconciliation following a fight, more effective at recruiting allies in fights, and more likely to participate in defending the group against a potentially intrusive animal. The same pattern of rank-related differences was shown by males. These data suggest that dominant animals differ from subordinate animals in their ability to maintain social bonds, to engage in appropriate aggressive behavior, and to inhibit inappropriate or unprovoked attacks. Combined with observations of other species, these findings strongly support the view that dominance and aggression are distinct concepts. They suggest that social skills, rather than physical prowess, are particularly important in the attainment and maintenance of high dominance rank.

In addition to underscoring the distinction between dominance and aggression, it is also important to emphasize that aggression is not merely a destructive social behavior: in group-living animals, aggression serves many functions. For instance, among vervet monkeys aggression is used to promote weaning and independence among immature members. Among juveniles, aggression is an important component of play. For adults, aggression may be used to control access to desired resources (such as shelter, shade, or sexual partners). Aggression is also important in the defense of a group's territorial boundaries from other vervet monkey groups. Further, under some circumstances, aggression may serve to protect animals from potential predators.

In view of its many functions, it is not surprising that many biological systems contribute to the mediation of aggression. For vervets and other monkeys, these include such morphological factors as large canine teeth and powerful temporal muscles, such endocrine processes as secretion of testosterone and other hormones, and such neurotransmitter systems as serotonin and norepinephrine. These systems are interrelated and bidirectional: for example, alterations in endocrine profiles (e.g., the relationship between estrogen and progesterone) influence brain serotonergic receptors (12), and, conversely, alterations in serotonergic function impact endocrine profiles. Because of the complexity of aggression and the multitude of biological and social systems that influence its expression, it is exceedingly unlikely that disruptions in serotonergic function are the most significant cause of destructive aggression. Nonetheless, there is substantial evidence implicating serotonergic systems in the genesis and expression of aggression.

PHYSIOLOGICAL RELATIONSHIPS

Among adult monkeys living in stable social groups, there is an inverse relationship between CSF 5-HIAA and the rate of aggression initiated. Figure 9.1 presents data

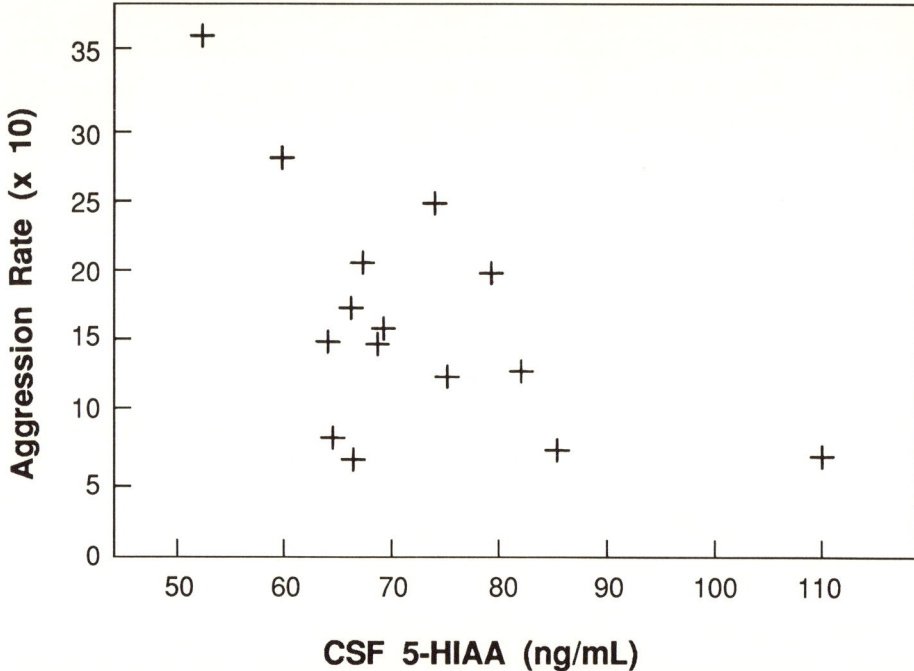

Figure 9.1. There is an inverse relationship between CSF 5-HIAA concentrations and the rate of initiating aggressive behavior in 14 subordinate adult males. Animals were members of stable social groups and their behavior was observed over a six month period. CSF 5-HIAA values represent each animal's mean concentration and are based on at least three determinations per animal.

from 14 subordinate adult males who were members of seven different social groups. Observed for four months, these males exhibited substantial interindividual differences in the rate of aggression. Similarly, there was approximately a threefold range in the CSF 5-HIAA. The figure clearly shows that among subordinate males, animals with low levels of CSF 5-HIAA were more likely to be aggressive than were those with higher CSF 5-HIAA concentrations.

When the influences of gender and rank on aggression are ignored, the relationship between reduced CSF 5-HIAA and heightened spontaneous aggression virtually disappears. When all animals are lumped together in a single group, irrespective of their gender and social status, the magnitude of the correlation between CSF 5-HIAA and aggression declines dramatically ($r = -.15$; n = 60; $P > .05$). There are consistent, albeit small, gender and status differences in basal CSF 5-HIAA. Resting CSF 5-HIAA is about 20% higher in females than in males and about 10% higher in dominant than in subordinate animals. However, there are substantial gender and status differences in the rate and intensity of aggression. Thus, it is reasonable to divide animals into four groups on the basis of status and gender and to ascertain whether, within each group, there is a link between CSF 5-HIAA and aggression.

Within each of the four groups of adult animals, there was an inverse correlation between CSF 5-HIAA and several measures of aggression (see table 9.3). For subordi-

Table 9.3. Status, Gender, Serotonin, and Aggressivity

Status	Gender (N)	Rate	Prolong	Escalate
Dominant	Female (12)	−.25	−.59*	−.61*
Dominant	Male (13)	−.28	−.63*	−.36*
Subordinate	Female (17)	−.66*	−.73*	−.81*
Subordinate	Male (18)	−.89*	−.91*	−.93*

Note: This table shows the effects of social status (dominant or subordinate) and gender on the relationship between serotonin and the rate, likelihood of prolonging, and likelihood of escalating aggression. Serotonergic function is inferred from each animal's concentration of CSF 5-HIAA. The number of subjects in each group is indicated in parentheses following the gender. Each cell represents the Pearson product-moment correlation between CSF 5-HIAA and the behavioral measures. An asterisk indicates $P < .05$.

nate females and for subordinate males, there was a robust inverse relationship between the rate of aggression and CSF 5-HIAA. The magnitude of this inverse relationship was reduced among dominant animals. Indeed, the correlations were only −.25 and −.28 for dominant females and dominant males, respectively. Neither correlation is significant. Because of the small numbers of subjects involved and the retrospective nature of this study, this apparent status-linked difference may be spurious. Nonetheless, it is possible that the aggression initiated by high-ranking animals is more likely to reinforce, rather than disrupt, social relationships, than is aggression initiated by low-ranking individuals. In that case, the absence of a link between the rate of aggression and diminished CSF 5-HIAA among dominant animals may reflect the adaptive, rather than the destructive, nature of some of the aggression initiated by dominant animals.

Other observations are compatible with this view. For example, within each of the four types of animals, CSF 5-HIAA is inversely related to the likelihood of prolonging a fight. This behavioral measure is not influenced by individual differences in the rate of aggression. Rather, it characterizes how likely it is that, once an animal is in a fight, that fight will be prolonged. Agonistic encounters are usually terminated by one animal submitting, avoiding, attempting to engage in grooming, or walking away. When one of these behaviors occurs, it is unusual for the other participant to continue to aggress. The occurrence of additional aggression represents prolongation of fighting. Dominant females and males were more than three times less likely to prolong fights than were subordinate animals (5% vs. 17%). Still, within each of the four types of animals, there were pronounced individual differences in the likelihood of prolonging a fight. And for a given type of animal, those individuals with the lowest CSF 5-HIAA concentrations were more likely to prolong fights.

Another parameter relating to the severity of aggression is the likelihood of escalating a fight. Escalation occurs when one of the participants in an agonistic encounter submits to or avoids a second participant and the second animal responds with more intense aggression. For instance, if A threatened B, B submitted, and A then bit B, A would be regarded as having escalated the encounter. Escalation is a probability measure that refers to the likelihood of an animal intensifying a fight given that a fight has already started. It is not related to the rate of fighting, and there were substantial gender- and status-linked differences in the rate at which animals escalated. Subordinate males were strikingly more likely to escalate fights than were other animals. However, within each class of animals, the probability of escalation was inversely related to CSF 5-HIAA.

PHARMACOLOGICAL STUDIES

We have also used pharmacological interventions to evaluate the roles of serotonergic systems in the control of destructive aggression. These studies examined the behavioral effects of drugs that exert their primary effect on serotonergic systems. Table 9.4 lists these compounds and the principle means through which they affect serotonergic systems. One set of drugs, which includes tryptophan, fluoxetine, and quipazine, enhances serotonergic function. Each compound relies on a different mechanism to enhance brain serotonin function.

In addition to its direct effect on serotonergic systems, each drug has a multitude of side effects. Tryptophan, for example, competes with other neutral aromatic amino acids for access to the carrier system that conveys it across the blood-brain barrier to enter the brain (13). One of these competing amino acids is tyrosine, the amino acid precursor to the catecholamines. Thus, tryptophan may indirectly alter catecholamine's function by decreasing the availability of the precursor. The other two compounds, fluoxetine and quipazine, may also exhibit many side effects. However, one effect common to all three drugs is the enhancement of serotonergic function. Thus, to the extent that they produce similar behavioral effects, it is parsimonious to assume that these commonalities result from increases in serotonergic function.

A similar rationale is involved in the interpretation of the effects of drugs that reduce serotonergic function. When given chronically, parachlorophenylalanine (PCPA), fenfluramine, and cyproheptadine decrease serotonergic activity, albeit by different

Table 9.4. Drugs Influencing Serotonergic Function

Compound	Polarity	Possible Mechanism
TR	+	Precursor
FL	+	Reuptake inhibitor
QP	+	Receptor agonist
PCPA	−	Inhibits synthesis
FEN	−	Disrupts storage
CYP	−	Receptor antagonist

Note: Tryptophan (TR), fluoxetine (FL), and quipazine (QP) are compounds that should increase serotonergic function. In contrast, chronic parachlorophenylalanine (PCPA), fenfluramine (FEN), and cyproheptadine (CYP) should diminish serotonergic function.

mechanisms (14). In addition to their impact on serotonergic systems, each of these compounds also affects other systems. Nonetheless, by parsimony, it can be hypothesized that the behavioral and endocrinological effects common to these treatments may be attributable to reductions in serotonergic function.

There are at least three criteria that we use to infer that serotonergic mechanisms are involved in the mediating of the behavioral effects of these drugs. These are illustrated in table 9.5, which shows the effects of selected drug treatments on spontaneous aggression directed toward other group members. One criterion is that the behavioral effects should be dosage dependent. Within limits, larger doses of a drug should result in larger, or more rapidly appearing, behavioral effects. Demonstration of a dosage-dependent relationship supports the view that the behavioral changes are proportional to the magnitude of the changes in the serotonergic function. Table 9.5 shows that increasing doses of tryptophan, fluoxetine, and quipazine resulted in increasing effects on spontaneous aggression.

A second criterion is that the direction of the behavioral effect produced by drugs that diminish serotonergic function (antagonists) should be opposite to that resulting from drugs that increase serotonergic function (agonists). Chronic fenfluramine, PCPA, and cyproheptadine increase spontaneous aggression. In contrast, tryptophan, fluoxetine, and quipazine decreased the rate and intensity of this behavior. In general, when assessing the role of serotonergic systems in the mediation of particular behaviors, the direction of the change in behavior is not critical because some behaviors (e.g., grooming) are increased by augmenting serotonergic function while others (e.g., vigilance and aggression) are decreased. What is important is that the polarity of the behavioral effects resulting from agonists and antagonists be reversed. This type of observation lends strong support to the view that serotonergic mechanisms are intimately involved in the mediation of a behavior.

The third criterion is that concurrent treatment with an agonist abolishes the effects of an antagonist (or vice versa). For example, concurrent treatment with quipazine abolishes the effects of cyproheptadine on aggression. As mentioned above, by itself

Table 9.5. Drug Effects on the Rate of Spontaneous Aggression

Drug	Dosage Level		
	Low	Middle	High
PCPA	131 ± 17	169 ± 18	226 ± 21
FEN	166 ± 19	245 ± 21	308 ± 29
CYP	212 ± 21	369 ± 31	410 ± 33
TR	85 ± 10	69 ± 8	51 ± 6
FL	79 ± 12	66 ± 7	48 ± 10
QP	81 ± 9	70 ± 10	56 ± 4

Note: Each drug was given in three doses to at least nine animals. The numbers represent the percentage of the rate (± standard error) of aggressive behavior relative to vehicle (baseline) levels.

cyproheptadine increases aggression, and by itself quipazine diminishes aggression. The observations that simultaneous treatment with both drugs eliminates the effects of either drug by itself lends further, strong support to this view.

Pharmacological alterations in serotonergic function also affect more subtle measures of aggressive behavior. Table 9.6 shows that, when treated with a drug that decreases serotonergic function, subjects are more likely to prolong and to escalate fights. In vehicle-treated (basal) conditions, subjects typically prolonged about 12% of their fights. When receiving a drug that reduces serotonergic function, the probability of an animal prolonging a fight doubled. In contrast, serotonin agonists substantially decreased the likelihood of prolonging fights. Relative to basal conditions, while receiving drugs that increased serotonergic function, animals were less than half as likely to prolong fights. That there were similar drug effects on escalating fights is also indicated. Reducing serotonergic function increased, and augmenting serotonergic function decreased the probability of fights becoming more severe.

The effects of drugs that alter aggression appear to be constrained by social status. Some time ago, we showed that dominant social status facilitated the behavioral effects of serotonergic agonists on prosocial behaviors (15). Relative to other males, dominant males responded to lower doses of tryptophan, fluoxetine, and quipazine. Further, at higher doses of these compounds, the magnitude of the behavioral changes was greater in dominant animals.

Social status is an important constraint on the effects of drugs that diminish serotonin function (see table 9.7). Preliminary data suggest that, relative to dominant animals, subordinate males are particularly at risk for the antisocial effects resulting from treatment with serotonergic antagonists. Small doses of drugs that diminish serotonergic function ignite aggressive behavior in subordinate animals but have no effect on dominant animals. At low dosages, subordinate animals not only show increases in the rates but also in the intensities of aggression. This type of aggressive behavior is exceedingly disruptive to the social group. Occasionally, subordinate animals treated with small doses of cyproheptadine were observed attacking and wounding juvenile group members. These attacks were not provoked by the juveniles and were immedi-

Table 9.6. Drug Effects on Prolonging and Escalating Aggression

	Agonist	Vehicle	Antagonist
Prolonging	4.6 ± 1.3*	11.9 ± 2.5	25.6 ± 3.9*
Escalating	8.2 ± 1.5*	21.3 ± 3.7	41.3 ± 6.8*

Note: This table shows the effects of serotonergic agonists, vehicle, and antagonists on the likelihood of prolonging and escalating aggression. Numbers in the cells represent the percentage of fights (± standard error) that are prolonged or escalated. The data in the agonist column are derived from separate studies in which subjects were treated with tryptophan, quipazine, or fluoxetine. The antagonist data are also pooled from separate studies in which subjects received PCPA, fenfluramine, or cyproheptadine. Each cell is based on the scores from at least 29 animals. An asterisk indicates $P < .05$, relative to vehicle (baseline) levels.

Table 9.7. Social Status Constrains the Effects of Serotonergic
Antagonists on Aggression

Behavior	Status	Dosage Level		
		Minimal	Moderate	Maximal
Rate	Sub.	310 ± 29*	433 ± 58*	621 ± 39*
	Dom.	102 ± 19	195 ± 31*	213 ± 42*
Prolong	Sub.	147 ± 20	182 ± 14*	270 ± 32*
	Dom.	110 ± 21	96 ± 13	108 ± 18
Escalate	Sub.	123 ± 16	208 ± 34*	326 ± 38*
	Dom.	95 ± 19	103 ± 22	106 ± 13

Note: This table shows the impact of social status on aggression induced by
treatment with serotonergic antagonists. The rate, the likelihood of prolonging,
and the likelihood of escalating aggression were recorded in subordinate (sub.)
and dominant (dom.) adult males. Each subject received a minimal, moderate,
or maximal dosage of at least one serotonergic antagonist. Data are in
percentage (± standard error) of a behavior relative to its rate in vehicle
(baseline) conditions. An asterisk indicates that the measure differs from vehicle
conditions for that group of animals.

ately followed by large-scale coalitions, led by the juveniles' mothers, against the
attacking subordinate animals. Subsequently, the treated animals persisted in aggress-
ing against adult males but refrained from attacking juveniles and infants.

Subordinate animals appear to be resistant to the ameliorative effects of concurrent
treatment with serotonergic agonists. For dominant individuals, a small dose of
concurrent quipazine treatment blocked the fenfluramine- and cyproheptadine-in-
duced increases in aggression. In contrast, subordinate individuals required a 10-fold
increase in dosage of quipazine to achieve a similar behavioral effect.

Admittedly, these investigations have been conducted in an exceedingly small
number of individuals; there is always the possibility that preliminary results may not
generalize to a larger sample of monkeys, let alone to human populations. Nonetheless, these observations suggest that subordinate social status may make animals
particularly vulnerable to the effects of even very small reductions in serotonergic
function. Such animals appear to be especially at risk for engaging in disruptive,
inappropriate aggressive behavior.

IMPLICATIONS

The data we have presented in this chapter show that, among adult monkeys living
in stable social groups, there is a robust link between diminished serotonergic function
and destructive aggression. Physiological investigations indicate that animals with
spontaneously low levels of CSF 5-HIAA appear to be predisposed to initiating,
prolonging, and intensifying aggression. The link between low levels of serotonergic

function and aggression is further supported by pharmacological studies, which demonstrate that drug-induced reductions in serotonin result in dramatic increases in severe, prolonged, and intense aggression.

This chapter also underscores the importance of social factors in constraining the link between serotonin and aggression. Adults with high social status may be shielded from engaging in drug-induced destructive, impulsive aggression. Dominant status may also protect adults with low basal CSF 5-HIAA from exhibiting spontaneous, maladaptive aggression. Thus, social and biological factors interact to render animals either at risk for, or protected from, engaging in destructive aggression.

These observations support the view that serotonergic systems play important inhibitory roles in confident, dominant animals. Rather than responding impulsively to social cues, dominant individuals seem to be able to time their aggressive behaviors so that social impact is maximized. In contrast, individuals with low serotonergic function may be inclined both to misread and to hyperrespond to social cues. In these disinhibited animals, innocuous social signals (e.g., juvenile play fighting) may be interpreted as unusually threatening and lead to the instigation of attack behavior.

In the coming decade there will be substantial insight into the genesis, maintenance, and consequences of violent criminal behavior. We anticipate that monkey studies may inform human investigations in at least three ways. Studies of nonhuman primates may provide data on the developmental courses that channel individuals into situations where they are likely to engage in violent, destructive aggression. A seminal set of macaque studies, summarized by Kraemer et al. (16) and by McKinney (17) has demonstrated that separating an infant from its mother or peers results in enduring alterations in monoaminergic function. As juveniles and young adults, animals with these histories are far more likely to engage in severe, frenzied aggression when challenged with amphetamine. Since separation may mimic many of the deficits in parental bonding and child care that some human infants encounter, Kraemer and McKinney's observations are compatible with the view that experiences during development may dramatically influence the ontogeny of the neural systems that regulate aggressive behavior in adults. One of our long-term goals is to extend these types of prospective studies to vervet monkeys. We expect to determine whether the links between physiological profiles and behavioral propensities shown by adult vervets have developmental precursors. For example, the link between CSF 5-HIAA and altered patterns of aggression may be present prior to its manifestation by adults. In such a situation, these animals may serve as appropriate models of humans at risk for initiating interpersonal violence.

A second arena in which monkey studies may contribute substantially to the interpretation of human behavior involves documenting the impact of social instability on the relationships between physiology and behavior. In many species, including humans, an unstable social milieu greatly exacerbates the likelihood of aggression. Among vervet monkeys, the introduction of an unknown conspecific or the removal of an established high-ranking leader from a social group often leads to heightened rates and altered patterns of aggression. During the three months following these types of alterations in group composition, the overall rates of wounding and other severe forms of aggression skyrocket and prosocial behavior declines. Nonetheless, there are striking individual differences in the animals' responses to the challenges posed by alterations in the composition of social groups. Preliminary data show that the bulk of the aggression and social disruption arises from the behavior of approximately one-quarter of the animals. Further, the extreme forms of aggression resulting in ostracism, wounding, and impaired reproductive function are present in only about 10% of the group members. It may be possible to utilize biological markers to identify individuals who are at risk for initiating and maintaining such intense, disruptive, aggressive behavior. These individuals may serve as an animal analogue to humans who have difficulty responding adaptively to alterations in social relationships.

Finally, a growing body of data indicates that genetic factors may also be causative agents in the etiology of impulsive aggression, some types of alcoholism, and other behavioral patterns leading to problems with the legal system. Preliminary data from Suomi and his colleagues' studies suggest that in rhesus macaques, individual differences in temperament may be in part genetically determined (18). Among infants and juveniles, these genetically based differences in temperament may account for many of the individual differences in the animals' behavioral and physiological responses to challenging situations. It has yet to be determined whether differences in temperament among developing animals contribute to individual differences in aggressivity and proclivities to violence exhibited among adolescent and adult animals.

Similarly, in free-ranging settings, a male vervet monkey migrates at adolescence from his natal group to another group. Some do not survive this process. In captivity, we simulate this transfer by moving adolescent males from their natal groups to other groups. This transfer is a difficult challenge, and, while no male has died in the process, approximately 40% (28 of 69 cases) do not become well integrated into their new groups. Thus, males may be divided into two cohorts on the basis of how adeptly they cope with a major life challenge. Retrospective studies have shown that, prior to transfer and during the first six months in their new groups, the cohort of males who have difficulty with the transfer process are more aggressive and less social than those who integrate easily into their new groups. In turn, membership in these cohorts may be predictive of subsequent behavior. Several males have been observed for three years following their initial transfer. Very preliminary observations suggest that males who transfer with great difficulty are likely to inappropriately aggress when confronted with social challenges. Thus, it may be that animals who have difficulty transferring out of their natal groups are at risk for engaging in maladaptive, explosive, aggressive behavior several years later. It would be interesting to determine if there are genetic factors that contribute in a causative way to these different patterns of adolescent aggressivity, migration, and subsequent propensity for violence.

In summary, a multitude of causative factors underlie the genesis, development, expression, and consequences both of adaptive, functional aggression and of destructive, inappropriate violence. The identification of biological factors that underlie complex social behavior in no way diminishes the importance of environmental and historical factors in shaping behavior. Understanding the complexity of biobehavioral relationships and using this information in a fashion that promotes a more humane and enlightened approach to legal issues remains a formidable challenge.

ACKNOWLEDGMENTS

This chapter describes the results of research that has been supported in part by the National Institutes of Health, the Veterans Administration, the Harry Frank Guggenheim Foundation, and the Elsie and Giles Mead Foundation. Competent intellectual and technical assistance has been provided by Deborah Bergin Pollack, Nuria Kimble, Glenville Morton, Jennifer White, David Torigoe, and Dan Diekman. Over the past decade we have benefited greatly from thoughtful discussions with Drs. Brammer, Dillon, Fairbanks, Heisel, Schuster, and Yuwiler.

REFERENCES

1) Ziegler, T.E., et al. Social interactions and determinants of ovulation in tamarins (*Saguinus*). In: *Socioendocrinology of Primate Reproduction*, T.E. Ziegler and F.B. Bercovitch (eds.), 113–133, Wiley-Liss, 1990.

2) Abbott, D.H. Behaviourally mediated suppression of reproduction in female primates. *J. Zool.* 213:455–470, 1987.

3) Worthman, C. Socioendocrinology: Key to a fundamental synergy. In: *Socioendocrinology of Primate Reproduction*, T.E. Ziegler and F.B. Bercovitch (eds.), 187–212, Wiley-Liss, 1990.

4) Raleigh, M.J., and McGuire, M.T. Social influences on endocrine function in male vervet monkeys. In: *Socioendocrinology of Primate Reproduction*, T.E. Ziegler and F.B. Bercovitch (eds.), 95–111, Wiley-Liss, 1990.

5) Miczek, K.A. The psychopharmacology of aggression. In: *Handbook of Psychopharmacology Vol.19: New Directions in Behavioral Pharmacology*, L.L. Iversen, S.D. Iversen, and S.H. Snyder (eds.), 183–328, Plenum Press, 1987.

6) Heninger, G.R., Charney, D.S., and Sternberg, D.E. Serotonergic function in depression: Prolactin response to intravenous tryptophan in depressed patients and healthy subjects. *Arch. Gen. Psychiatry* 41:398–402, 1984.

7) Brownstein, M.J. Serotonin, histamine, and the purines. In: *Basic Neurochemistry*, G.J. Siegel, R.W. Albers, B.W. Agranoff, and R. Katzman (eds.), 219–231, Little, Brown, and Co., 1981.

8) Yuwiler, A., Brammer, G.L., Morley, J., et al. Short-term and repetitive administration of oral tryptophan in normal men. *Arch. Gen. Psychiatry* 38:619–626, 1981.

9) Roy, A., Virkkunen, M., Guthrie, S., and Linnoila, M. Indices of serotonin and glucose metabolism in violent offenders, arsonists, and alcoholics. *Ann. NY Acad. Sci.* 487:202–220, 1986.

10) Parnavelas, J.G., and Papadopoulos, G.C. The monoaminergic innervation of the cerebral cortex is not diffuse and nonspecific. *Trends in Neurosciences* 12(9): 315–319, 1989.

11) Smuts, B.B. Gender, aggression, and influence. In: *Primate Societies*, B.B. Smuts, D.L. Cheney, R.M. Seyfarth, R.W. Wrangham, and T.T. Struhsaker (eds.), 400–412, University of Chicago Press, 1987.

12) Soubrie, P. Reconciling the role of central serotonin neurons in human and animal behavior. *Behav. Brain Sci.* 9:319–364, 1986.

13) Leathwood, P.D., and Fernstrom, J.D. Effects of an oral tryptophan/carbohydrate load on tryptoph, large neutral amino acid, and serotonin and 5-hyroxyindoleacetic acid levels in monkey brain. *J. Neural Trans.* 79:25–34, 1990.

14) Miczek, K.A., and Donat, P. Brain 5-HT and inhibition of aggressive behaviour. In: *Behavioural Pharmacology of 5-HT*, P. Bevan, A.R. Cools, and T. Archer (eds.), 117–144, Lawrence Erlbaum, 1989.

15) Raleigh, M.J., Brammer, G.L., McGuire, M.T., and Yuwiler, A. Dominant social status facilitates the behavioral effects of serotonergic agonists. *Brain Res.* 348:274–282, 1985.

16) Kraemer, G.W., Ebert, M.H., Schmidt, D.E., and McKinney, W.T. A longitudinal study of the effects of different social rearing conditions on cerebrospinal fluid norepinephrine and biogenic amine metabolites in rhesus monkeys. *Neuropsychopharmacology* 2(3):175–189, 1989.

17) McKinney, W.T. Interdisciplinary animal research and its relevance to psychiatry. In: *Comprehensive Textbook of Psychiatry*, V.H. Kaplan and B. Sadock (eds.), 326–339, Williams and Wilkins, 1989.

18) Higley, J.D., and Suomi, S.J. Temperamental reactivity in non-human primates. In: *Temperament in Childhood*, G.A. Kohnstamm, J.E. Bates, and M.K. Rothbart (eds.), 153–167, John Wiley and Sons, 1989.

10

Serotonin and Social Rank among Human Males

Douglas Madsen

Abstract: For humans, as among the vervets studied by Raleigh and McGuire (chapter 9, this volume), laboratory studies reveal a very strong link between social rank and whole-blood serotonin. In humans, these differences seem to be related to psychological or personality traits. For aggressive competitors—who have been called "Machiavellians" by psychologists—the relationship between serotonin and social rank is strongly positive. For the more deferent individuals, described as "moralists," the relationship is almost perfectly negative.

In this chapter I focus on a topic of continuing fascination to students of political life: the special nature of those individuals who seek, and sometimes acquire, positions of influence. I think most would agree that this is a subject in which fascination has not yet yielded much in the way of scientific understanding. In spite of a wealth of speculation about personality factors and social-background contingencies, we still know relatively little about why some people are influential and others are not. In what follows, I present remarkable, though still somewhat mysterious, findings on one aspect of that question.

In papers published earlier, I have provided evidence suggesting that social competitiveness, and possibly also social rank, may have a specific biochemical marker—namely, whole-blood serotonin (Madsen, 1985, 1986).[1] This initial result was broadly in keeping with very strong findings connecting elevated levels of whole-blood serotonin (WBS) to social dominance in a lower primate species.[2] My evidence, derived from a quasi-experimental study involving 72 undergraduate males, was of two types. The first was cross-sectional correlations between WBS and four defining elements of the type A behavior pattern: drive, aggressiveness, competitiveness, and mistrust of others.[3] The last of the four (now seen as the type A element involved in coronary disease) was a particularly intriguing member of this set of dispositions, and in the first paper I discussed it as a possible indicator of the psychopolitical disposition labeled Machiavellianism. Unfortunately, in the initial study lacked the data needed for further examination of this possibility.

The second type of evidence involved comparisons of time-series observations for several blood-borne "stress" hormones, collected while my 72 subjects were engaged in an ongoing competition. The argument that guided the analysis of the stress data was simple: if high-serotonin individuals are truly distinctive in their approach to competition, as suggested in the cross-sectional correlations, such distinctiveness might be observed in their psychoendocrine response to rivalry. This hypothesis was well supported by the data. Those with high WBS exhibited an internal response to competition which set them apart from "normals," especially in terms of blood cortisol levels.

Needless to say, these two papers left many more questions than answers. And of those questions, an especially important one for social science was whether WBS was actually tied to social rank, as with the aforementioned primates, or was merely tied to some psychological tools useful in the pursuit of rank. This question could not be given even preliminary examination in the first study because of the absence of any measures focused upon the social arenas from which the participants had come. Neither objective nor subjective assessments of the social rank of my subjects had been included.

To address this question and others, a second laboratory investigation was undertaken. After describing its methods, I present some findings from that investigation. These findings greatly strengthen but also complicate the earlier argument tying WBS to dominance sought and achieved.

METHODS

The investigation was housed at the Clinical Research Center of the University of Iowa Hospitals. Twenty subjects were selected from an applicant pool of 35 undergraduate males. The participants had been recruited by means of an advertisement stating that they would be competing in a tournament involving computer games, that blood samples would be drawn periodically during the session, that each participant would be paid $25, and that the winning player would get $200.

On his assigned day, each subject came to the hospital at 4 P.M. and had an indwelling catheter placed in his arm. The first blood sample was taken immediately, after which the subject completed a questionnaire treating various self-perceptions—in particular, perceptions of type A behavior, of risk orientation, of manipulativeness, of general efficacy, and of social rank. Each of the 84 questionnaire items was presented as an agree-disagree proposition, and the intensity of agreement or disagreement was registered by drawing a line across the paper (the maximum length being 34 cm).[4]

After completing the questionnaire, each subject ate a standard dinner, whereupon another blood sample was taken. He then moved to the room where the game would be played and was given the first briefing sheet. This sheet presented a set of situations in which one actor confronted another and could opt for either an aggressive or a deferent course of action. Each of these situations represented a "chicken" game, and the subject was asked to think about the circumstances in which it was better to rise to a challenge and those in which it was better to walk away.[5]

Ten minutes later, a second briefing sheet was presented, this one giving some particulars of the game being used in the study. This sheet reminded the subject that his opponent was to be a "smart" computer program and went on to say:

> In reality, of course, this means you are playing against a smart computer *programmer*. This person simply has laid out his strategies within a computer program. In the program, he tries to anticipate your moves, and he writes instructions for how your tactics are to be countered. He accumulates information on your moves and then looks for a pattern in your decisions. He reacts to what you do and changes his approach where needed. He learns from his mistakes. In sum, he has set up a strategic game against you, just as can be done in computer programs that play very good chess. His goal, like yours, is to win.
>
> This program represents the best that the programmer, an accomplished gamer, can throw at you. . . . For you to do well, you must figure out what he is up to—what tactics he is using—while the contest unfolds. Obviously, intelligence counts here.

The second briefing underscored the point that the programmer was making his moves in the dark (i.e., that his choices were established before the subject made a move). Only past behavior would be known by both sides. It was also emphasized that the game was open-ended; it would last well over an hour, but one could not predict just when the last move would come. Lastly, it was explained that, in making each move, he must think aloud about the choice being made.[6] The pros and cons of his alternatives were to be discussed and the current situation reviewed out loud.

At this juncture, the player was joined by a monitor, who remained present throughout the game. The monitor started the audio recording and led his subject through some practice in thinking aloud (while solving simple anagrams and a multiplication problem and while reflecting upon an example of a chicken game). Then, and each 20 minutes thereafter, blood samples were taken.

The third and final briefing sheet was presented next. It described the simple scenario for the game: two countries, having a long history of conflict, must decide whether to mobilize forces against each other. The four-cell payoff matrix (which was shown graphically in the form of a table left with every subject throughout play) stipulated the following: a player would gain or lose no points if neither he nor the computer took aggressive action and mobilized; there would be a loss for each of 350 points if both mobilized; and in the asymmetrical case, the change would be +175 for the mobilizer and −175 for the nonmobilizer. The monitor made sure that the payoff system was well understood by each subject.

The game started at 6 P.M. and ran until 7:20. After each move, the player was prompted on screen to speculate aloud about what his opponent was trying to do and what he should now do. After each move, the new scores for the player and his "opponent" were displayed. After every 10 moves, the player was prompted on screen to state aloud how well he thought he was doing, with the evaluation scale running from 1 to 10. Throughout the contest, each player spoke steadily about his moves and his opponent's moves. Although the monitor was present primarily to remind the player to continue thinking aloud, it turned out that such prompting was seldom needed. Not surprisingly, the players dealt with the computer in an anthropomorphic fashion. The game was structured so that a player could not decipher his computer opponent's strategy, but at the end most players thought they had seen through at least part of that strategy. The behavior of the players suggested that they were deeply involved in the game, an impression that was confirmed by the players themselves in the postgame interviews.

The computer program specified that, whatever the first move by a player, the computer response always would be aggressive (that is, mobilization). Bear this point in mind. In what follows, I treat only the first two moves by each player. For those two moves, all players had experienced the same competitive situation; all had encountered precisely the same aggressive response from the opponent (though with different payoff implications).

At the end of the game, the monitor asked each player how he thought he had done, whether he had figured out the computer's strategies, whether the computer seemed to have figured out *his* strategies, whether the game was irritating or tension inducing, and whether he would be willing to play again.

Data generated by this investigation include the serotonin blood values, the time-series blood values for various stress hormones, the questionnaire-elicited self-assessments, the record of moves in the game, and a rather full text of spoken thoughts and evaluations recorded during play.

In both spirit and method, this was an exploratory exercise.[7] Although the central concern here (i.e., the relationship between WBS and social rank) was clear, the precise ways in which that concern should be addressed were less so. Plainly, I was not dealing with social rank in any larger sense. It was the intimate social arenas of everyday life that were in point (as had been the case with the primate studies). How these smaller units nest within greater ones and the implications of such nestings (if sensed or understood by those involved) for behavior and psychophysiology could only be matters of speculation.

Moreover, the obvious distinction must be made between objective and subjective rank. There is good reason to suppose that the latter would be more important for most psychophysiological responses. Objective rank *without* subjective appreciation of that rank, though possible, would not seem a likely condition for driving relevant internal processes; oddly mixed signals would be ubiquitous, and confusion probably would reign. However, even were that not so, it simply was not feasible for me to have established empirically, in the fashion of the human ethologist, just what the objective social ranks of my research subjects might be. Such would have required a separate and very complex investigation for each and every player in the contest. And the investigation would have to have occurred within the social arena(s) *that the subjects themselves believed most important.*

However, if necessity and preference converged here in measuring social rank, that did not set aside the perennial problem of reliability in self-assessment. In establishing subjective social rank, I used four questionnaire items.

> •I am a leader in my social circle of friends.
> •I am more likely to be a leader than a follower.
> •Many of my friends look to me for guidance.
> •In my social circles, I am pretty much at the top these days.

These items were widely separated from one another in the questionnaire. They were not, recall, items for which only simple-answer categories were provided. Instead, magnitude scaling (using line production) was employed to measure the conviction of each response. Thus, it was satisfying to find that these items had a mean intercorrelation of .73, and that Cronbach's alpha coefficient for the four was .92. By all conventional standards, these items make up a very reliable measure. It is reasonable, then, to assume that the subjects believed what they said about their own social status. The accuracy of those beliefs, however, was untested.

FINDINGS

In the first test of the relationship, I found that none of the four leadership items was related to WBS in a simple bivariate analysis. The correlation coefficients hovered around zero, and the scatter diagrams displayed what seemed to be patternless data points. This clearly was not what the primate model had seemed to predict.

I turned to how the game was played. (Because of a single lapse in a recording program, this reduced my cases by one.) Were the aggressors on the first move—when all else about the game itself was still equal—different, either in terms of self-assessed social rank or in terms of WBS? In the simple bivariate test they were not.

Only when the three variables—WBS, perceived social rank, and game play—were considered jointly did extraordinary results emerge.[8] Figure 10.1 presents them.

Looking past the single outlier in each of the top two diagrams, one finds patterns standing out quite clearly; indeed, in the case of the deferent group, astonishingly clearly. (Two of the data points for this group actually coincide.) For the deferent group, the WBS relationship with social rank is strongly negative; those higher in rank have lower WBS and vice versa. For the aggressors, however, the relationship is completely different; those high in rank have higher serotonin levels, those low in rank have lower.[9]

The scatter diagram for the total group, also presented, shows how these cross patterns produced a null relationship overall. (And it shows, intriguingly, that the outlier in the aggressive group falls directly on the regression line for the deferent group.)

Robust regression analysis can be used to summarize these relationships.[10] Note that in this analysis social rank is treated as the dependent variable (an especially important choice when using these particular robust methods) because of its clearly dependent status in the lower primates. Summary results are given in table 10.1.

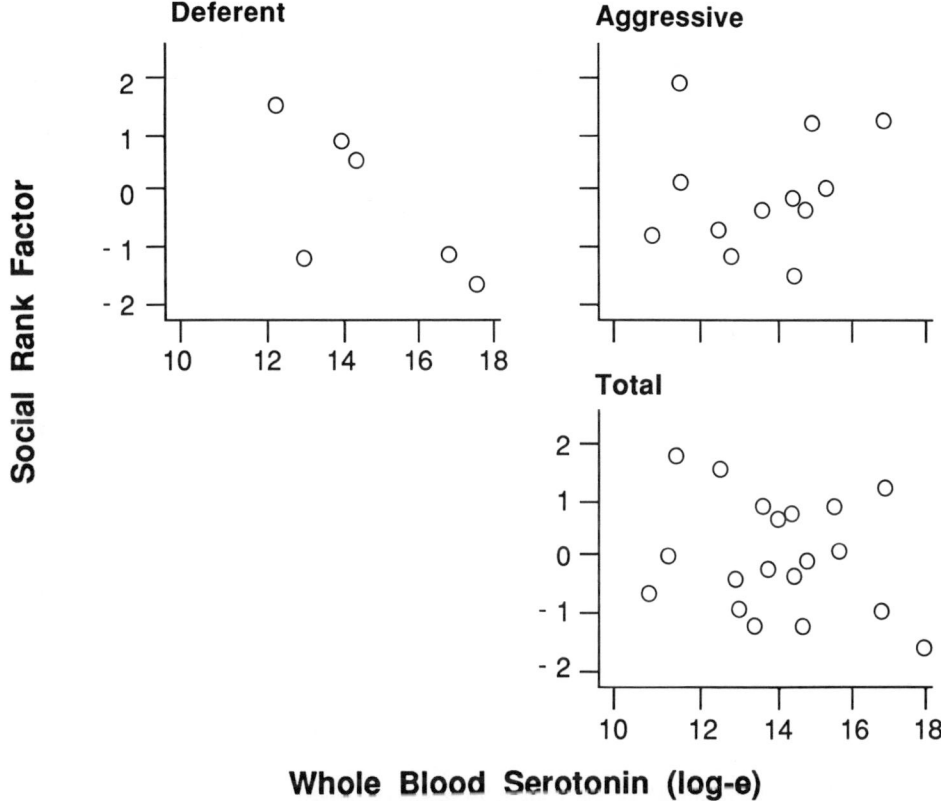

Figure 10.1. Social rank and WBS, by first move.

If controlling for the first game move reveals a dramatic relationship between social rank and blood serotonin, what might be added by considering the second move? Remember that, no matter what the first move by a player, the computer response was aggressive (to mobilize). Hence, after that first move the deferent player saw that he had lost 175 points and his opponent had gained 175 points. And the aggressive opponent saw that both he and his opponent had lost 350 points. (Again, to be sure no confusion arose about the result of each move, the changing point counts were presented on screen.) Of the 12 players who took the aggressive path on the first move, nine stuck with that choice on the second. Of the seven who deferred on the first move, six were brought to the aggressive stance on the second. Thus, there were players in each of the four possible categories after the second move. The scatter diagram for each is given in figure 10.2.

The single player who was deferent in both moves now can be seen to have been the outlier in the nonaggressive group in move one. As a unique case he must here remain a mystery. Three cases went from aggressive to deferent, and little can be gleaned from this tiny set of observations. On the other hand, the defer-to-aggress group exhibits the relationship between social rank and WBS more clearly than ever; indeed, it is virtually perfect. And the aggress-aggress group also shows a clearer pattern, although still with one outlier.

Table 10.1. Robust Regression of Social Rank on WBS, by First Type of Move

	Constant	Slope	t	Prob	n	Adj R2
Deferent	8.8	−.58	−23.0	0.00	7	.989
Aggressive	−4.1	.28	1.8	0.10	12	.170

Regression summaries (for what I have labeled the def-agg and agg-agg groups) are given in table 10.2. The model for the former, as might have been expected, changes scarcely at all from that for the first-move-deferent group. The model for the latter, however, gains considerable statistical strength, in spite of its reduced number of observations.

These plainly are strong findings, but what do they mean? Why does the WBS effect reverse for the two key groups? What types of individuals populate these two groups? Fortunately, this last question could be addressed empirically. Turning to the questionnaire data, I found the two groups to be strikingly different on a number of psychological dispositions. The data are presented in table 10.3.[11] The student of

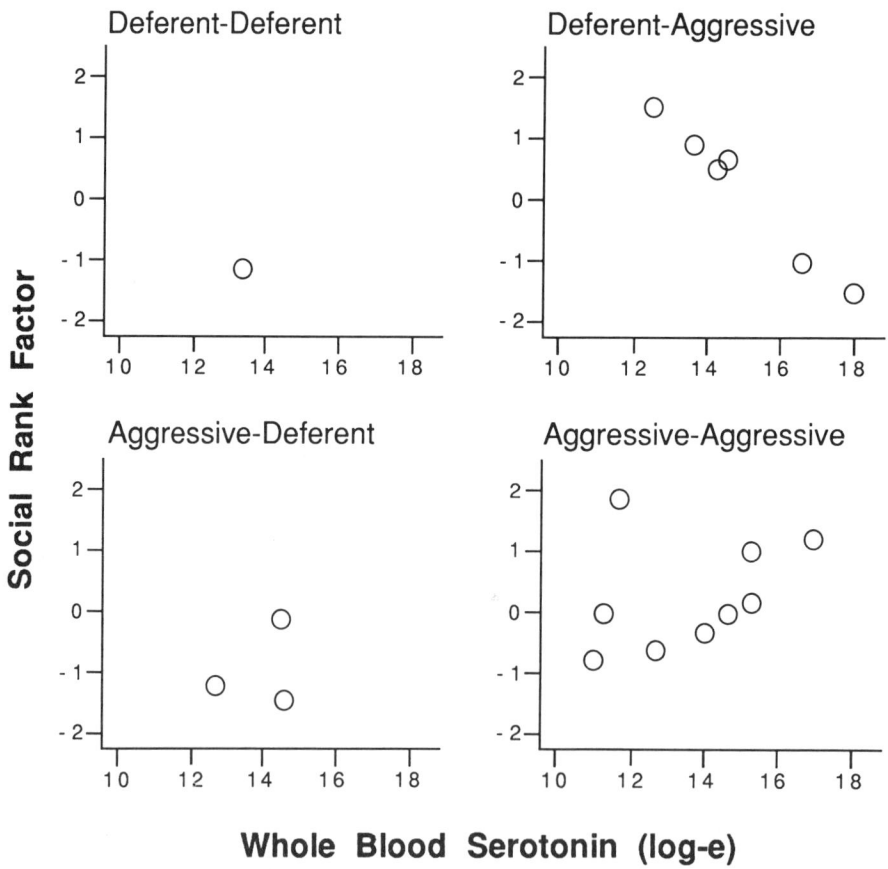

Figure 10.2. Social rank and WBS, by moves 1–2.

Table 10.2. Regression of Social Rank on WBS, by Two Moves

Move Pattern	Constant	Slope	t	Prob	n	Adj R2
Def-Agg*	8.8	−.58	−25.4	0.00	6	.992
Agg-Agg**	−4.1	.30	2.4	0.05	9	.369

* OLS estimation.
**Robust estimation.

political psychology will recognize that most of these differentiating attitude items come from the Machiavellianism scale developed by Christie and his colleagues (Christie and Geis, 1970).

These items were factor analyzed, and the first principal component, which captured 33% of the variance in the set of variables, was retained. That factor was dominated by item 12 (loading at .8); items 7, 13, 18, and 26 (loading at .7); and items 9 and 15 (loading at .6). In every case, the direction of the loadings was that of the correlations given for the aggressive group in table 10.3.

This factor pattern represents with striking clarity what Christie (1970) called the

Table 10.3. Correlations of Group Type and Differentiating Attitudes

Agg-Agg	Def-Agg	Questionnaire Location and Content of Attitude Item
−.406	+.371	7. One should take action only when sure it is morally right.
+.364	−.514	9. Sometimes people have to get hurt if important things are going to get done.
−.452	+.369	12. It's always best to reveal your real reasons if you want to get cooperation from people.
+.482	−.386	13. Most people are self-centered.
+.647	−.256	15. Life in the "fast lane" sounds great.
−.292	+.456	18. There is no excuse for lying to someone else.
−.345	+.496	25. People can be trusted.
+.421	−.107	26. Sometimes we all have to cheat a little to get what we want.
+.501	−.301	31. It is always better to go for the win than to settle for a tie.
+.426	−.107	36. People who can't make up their minds are a pain.
−.555	+.191	65. I am not very resistant to the influence of others around me.
+.378	−.375	70. I don't let social competition bother me.
−.469	+.302	81. Often I enjoy just sitting around and thinking.
−.433	+.295	82. Often I feel worried.

Note: Although it is the overall pattern here which is of primary importance, significance levels may interest some readers. With such a small number of cases, the .01 level of confidence is not attained until a coefficient reaches a very hefty .58. Correspondingly, for $P < .05$, it must reach .45; for $P < .15$, .35; and for $P < .20$, .31.

"duplicity" component of Machiavellianism.[12] At the top, we have those who can aptly be called manipulators: action oriented, focused on self-interest, willing to cut corners, casual about rectitude. At the bottom, we have moralists.

A graphic depiction of the position of each type of game player on this measure is given in figure 10.3. The graph presents box plots, a relatively new, and for some perhaps unfamiliar, display developed by Tukey (1977). In this type of plot, a distribution is shown as a box extending from the first quartile to the third. Within the box, a horizontal line shows the median value. Vertical lines extend from the top and bottom of the box to the last data points that lie within 1.5 times the interquartile range. Outliers are shown as individual dots. (Because of insufficient cases, the player who deferred twice is graphed as a dot in figure 10.3, and those who were first aggressive and then deferent are graphed as a simple box.) In figure 10.3, one can see that the duplicity measure rather neatly separates the two groups of size (the agg-agg and def-agg groups). There is almost no overlap. (And it is interesting to have the small agg-def group fall in an intermediate location here.) Thus, we have a stark contrast between the most aggressive players, largely manipulators, and the more peaceful players, largely moralists.

DISCUSSION

These results yield a complicated picture. In the earlier work on male vervets, the link between dominance and WBS was relatively straightforward. The dominant male had significantly elevated WBS; WBS levels rose and fell with the ascent to or descent from dominance; and, in an experimental test, the injection of a serotonin agonist produced dominance for the treated animal. Of course, there were many differences between this primate social system and that of humans; for example, dominance for the vervets was a binary condition: one male had it, and the others were of comparably low rank. In addition, the primate system consisted of a single social arena rather than a nested or multiple-arena system like that experienced by most urbanite humans.

It also may be important that the vervet studies dealt with actual dominance rather than with self-assessed rank. The possibility of such self-assessment being colored, consciously or unconsciously, by unmeasured psychological "contaminants" cannot be dismissed out of hand. Direct observation of social rank in humans, difficult though it is for any significant number of cases, still would be valuable.

Finally, even were all of the above differences set aside, there obviously is no way to translate the distinction between manipulators and moralists into something comparable for vervets. So what do we have here? Even with the small number of observations in this study, the extraordinary strength of the statistical findings seems to me to warrant three statements.

1. In humans there appears to be a remarkable connection between blood serotonin and social rank, certainly when rank is self-assessed and very possibly when it is objectively assessed as well. At the very least, this connection applies to young adult males, but it may be much more general than that.

2. The social-rank–WBS relationship is wholly different for aggressive individuals and for

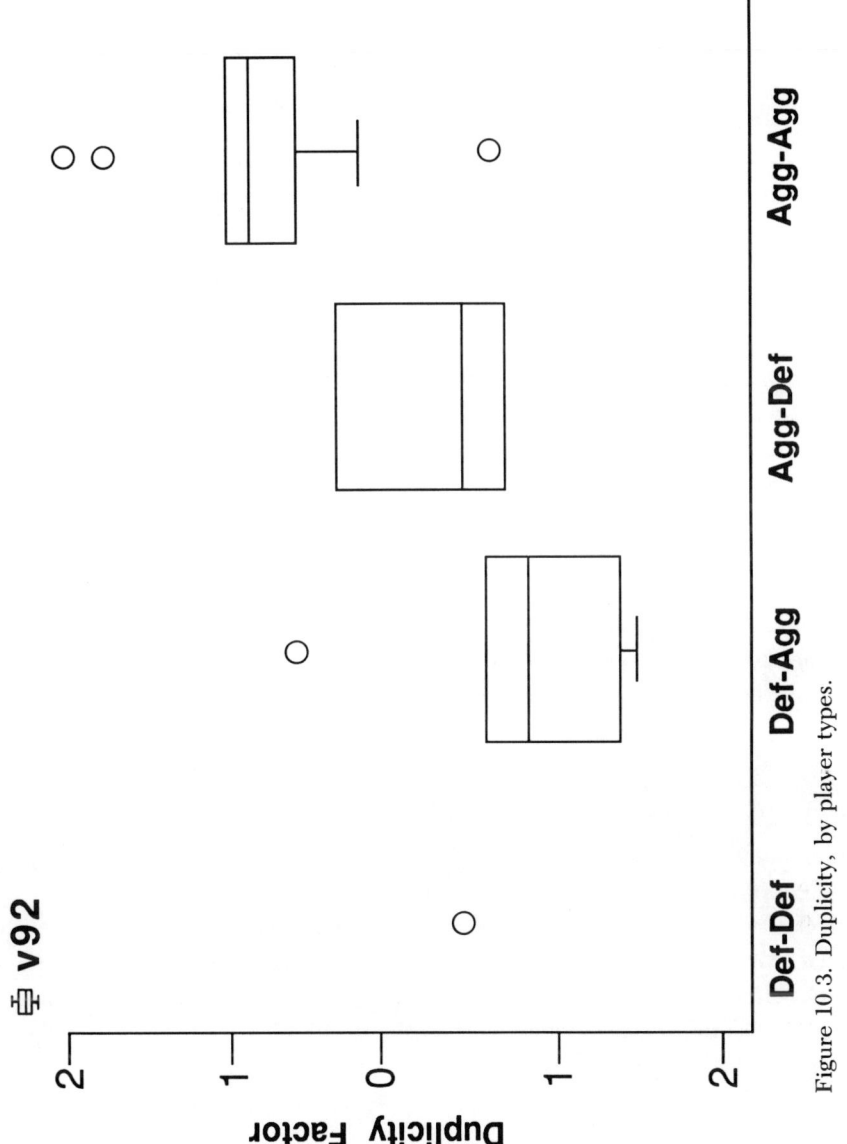

Figure 10.3. Duplicity, by player types.

155

nonaggressive individuals. For the aggressives, serotonin rises with rank. For the nonaggressives, it declines. Thus, in apparent contrast with the vervets, a higher level of WBS marks higher rank only among aggressives.

3. The aggressives themselves are a fascinating group. Seemingly classic Machiavellians, they provide a close fit with the politician of popular imagination: ambitious, self-centered, and when it comes to moral questions, a little fast and loose. If the aggressive-manipulative style enhances the likelihood of real-world political success (which is, of course, unproven), then the dominant figures in our major social arenas may in fact show the elevated blood serotonin predicted by the animal model.

Three large questions immediately follow from these results. First, in the course of replicating these findings, how far can they be extended? Can the results be reproduced for different age groups? For females? For objectively assessed leaders? For different cultures? Second, if WBS is a marker of social rank, how does it work? What drives WBS levels? What other psychological dispositions or neurochemical factors may be involved?[13] And third, how does the manipulative-moralistic dimension play into the acquisition of influence and rank? Is the Machiavellian disposition inherent to aggressiveness (of the kind measured here)? Or do aggressives sometimes display the moralistic bent?

Obviously, there are more questions than answers. Some of them are best left to the neuroscientists. However, as Harold Dwight Lasswell pointed out long ago, when it comes to studying influence and the influential, many of the most interesting questions lie squarely in the domain of political science.

ACKNOWLEDGMENTS

This research was supported by the Harry Frank Guggenheim Foundation and the National Institutes of Health. Other crucial contributions came from Dr. Janet Schlechte and her staff at the University of Iowa Clinical Research Center, from Dr. Michael McGuire of the UCLA Department of Psychiatry, from the late Dr. Diana Van Orden of the University of Iowa Department of Obstetrics and Gynecology, and from Ms. Joanne Madsen. I am grateful to one and all.

NOTES

1. I had assumed that no one could so completely misunderstand this work on status markers as to see in it some variant of biological determinism. That assumption turned out to be wrong. Hence, it seems I must belabor the obvious: there is every reason to believe that such markers respond to environmental circumstance; there is no reason to believe that they alone determine social rank or behavior.

2. In my earlier papers, I have cited many of the studies coming from Michael T. McGuire and his colleagues at UCLA—the team who discovered and then explored the tie between WBS and social dominance in male vervets. Some of their more recent work is found in Raleigh et al. (1985); McGuire et al. (1986); Brammer et al. (1987); and Raleigh and McGuire (1989). For a study connecting WBS to dominance in a second monkey species, see Steklis et al. (1986).

3. Note that, since WBS levels change gradually (i.e., over the course of days rather than

minutes or hours), the WBS values observed would not reflect the successes or failures experienced during the few hours of my study. For a recent report treating short-term WBS stability, see Kremer et al (1990).

4. This measurement procedure basically follows Lodge and Tursky (1981).

5. For a discussion of chicken games, characterized as "brinkmanship," see Schelling (1966).

6. The procedures used here were adapted from Ericsson and Simon (1984).

7. In recent years, there has been much greater (though far from unanimous) acknowledgment of the usefulness of exploratory data analysis in science. For a collection of papers on exploratory analysis, see Hoaglin et al. (1985).

8. Because of the high intercorrelations between the four questionnaire items on social rank, I extracted a single principal component from these four items. That factor is the measure of social rank to be used in what follows. This, and all other statistical analyses, were done within STATA. Note that the WBS measure has been logged to normalize its distribution.

9. One might think that the game move should not be characterized as either deferent or aggressive, as here, but rather as cooperative or selfish. I rejected that characterization because I found no difference between the two groups on questionnaire items that might be thought to tap such a dimension. For example, there was no response differential to "The old expression is right: It is a dog-eat-dog world." Nor was there one on "Success for oneself should be achieved before taking on the problems of other people."

10. As can be seen in the scatter diagrams, the data here present the classic case for robust, as opposed to ordinary, regression analysis. In each diagram, a single data point is set apart from the pattern established by the others and would, in ordinary regression analysis, exert unduly great and perhaps crippling influence on estimates. Robust regression treats such an outlier as an anomaly, which should not carry much weight in the estimates. It involves repeated application of OLS, with iterative weighting of the data on the basis of the residuals from the previous estimates, until the weights stabilize. The robust technique in STATA is that suggested by Guoying Li (1985), with Tukey's tuning parameter set to 7. On robust techniques in general, see Hoaglin et al. (1983, 1985) and Rousseeuw and LeRoy (1987).

11. Let me hasten to add that these differences are *not* a function of dissimilarity on the social-rank distributions for the two groups; those distributions are very similar.

12. The pattern also looks remarkably like Eysenck's "T" (for tough-mindedness) disposition, which he saw as having manipulators and force users at one extreme, with moralists and thinkers at the other; see Eysenck (1970).

13. It may be of interest that a rather exhaustive analysis of my data has failed to implicate either norepinephrine level or psychological orientation toward risk in the findings presented above.

REFERENCES

Brammer, Gary L., Michael T. McGuire, and Michael J. Raleigh. (1987). "Similarity of 5-HT2 receptor sites in dominant and subordinate vervet monkeys." *Pharmacology, Biochemistry and Behavior,* 27:701–705.

Christie, R., and F. Geis. (1970). *Studies in Machiavellianism.* New York: Academic Press.

Diaconis, Persi. (1985). "Theories of data analysis." In David C. Hoaglin, Frederick Mosteller, and John W. Tukey (eds.), *Exploring data tables, trends, and shapes.* New York: Wiley, 1–36.

Ericsson, K. A., and Herbert A. Simon. (1984). *Protocol analysis: Verbal reports as data.* Cambridge, MA: MIT Press.

Eysenck, H. J. (1970). *The structure of human personality.* London: Methuen.

Hoaglin, David C., Frederick Mosteller, and John W. Tukey (eds.). (1983). *Understanding robust and exploratory data analysis.* New York: Wiley.

————. (1985). *Exploring data tables, trends, and shapes.* New York: Wiley.

Kremer, H. P., Jaap G. Goekoop, and Godfried M. Van Kempen. (1990). "Clinical use of the determination of serotonin in whole blood." *Journal of Clinical Psychopharmacology,* 10:83–87.

Li, Guoying. (1985). "Robust Regression." In David C. Hoaglin, Frederick Mosteller, and John W. Tukey (eds.). *Exploring data tables, trends, and shapes.* New York: Wiley, 281–344.

Lodge, Milton, and Bernard Tursky. (1981). "On the magnitude scaling of political opinion in survey research." *American Journal of Political Science,* 25:376–419.

McGuire, Michael T., Gary L. Brammer, and Michael J. Raleigh. (1986). "Resting cortisol levels and the emergence of dominant status among male vervet monkeys." *Hormones and Behavior,* 20:106–117.

Madsen, Douglas. (1985). "A biological property relating to power-seeking in humans." *American Political Science Review,* 79:448–57.

————. (1986). "Power seekers are different: Further biochemical evidence." *American Political Science Review,* 80:261–269.

Raleigh, Michael J., and Michael T. McGuire. (1986). "Animal analogues of ostracism: Biological mechanisms and social consequences." Special Issue, Ostracism: A social and biological phenomenon. *Ethology and Sociobiology,* 7:201–214.

————. (1989). "Female influences on male dominance acquisition in captive vervet monkeys, *Cercopithecus aethiops sabaeus.*" *Animal Behaviour,* 38:59–67.

Raleigh, Michael J., Gary L. Brammer, Edward R. Ritvo, Edward Geller, et al. (1986). "Effects of chronic fenfluramine on blood serotonin, cerebrospinal fluid metabolites, and behavior in monkeys." *Psychopharmacology,* 90:503–508.

Raleigh, Michael J., Gary L. Brammer, Michael T. McGuire, and Arthur Yuwiler. (1985). "Dominant social status facilitates the behavioral effects of serotonergic agonists." *Brain Research,* 348:274–282.

Rousseeuw, P., and A. LeRoy. (1987). *Robust regression and outlier detection.* New York: Wiley.

Schelling, T. C. (1966). *Arms and influence.* New Haven: Yale University Press.

Steklis, Horst D., Michael J. Raleigh, Arthur S. Kling, and Ken Tachiki. (1986). "Biochemical and hormonal correlates of dominance and social behavior in all male groups of squirrel monkeys (*Saimiri sciureus*)." *American Journal of Primatology,* 11:133–145.

Tukey, John W. (1977). *Exploratory data analysis.* Reading, Mass.: Addison-Wesley.

Part 4

The Challenge to Legal Concepts

Does the discovery of biological factors controlling human social behavior—and sometimes contributing to criminal acts—require a change in the traditional concepts on which our law rests?

11

The Brain, the Law, and the Medicalization of Crime

C. Ray Jeffery

Abstract: Should crime be viewed as a condition requiring medical treatment rather than as an act deserving punishment? Models of criminal law in our tradition, whether based on the concept of rehabilitation or that of retribution, seem to have failed to achieve their stated objective. Neuroscientific research like that described in this volume suggests the propriety of turning to a medical approach to crime.

THEORETICAL ISSUES

Legal traditions grew out of philosophical idealism and rationalism, whereas scientific traditions grew out of philosophical empiricism and positivism (Jenkins, 1984; Jones, 1986; Mannheim, 1972; Rennie, 1978). Criminal law is based on concepts of free will, moral responsibility, punishment, and mentalism. Criminals must be punished because they are bad, not mad. A criminal must possess an evil mind, the legal requirement of mens rea. According to this philosophy of mind-body dualism, the mind controls the act (the actus reus); that is, there are nonphysical causes for physical events. If people behave badly it is because of free will and lack of moral responsibility, rather than because of biological, psychological, and social factors. If people wanted to, they could behave as law-abiding citizens. Our legal and ethical philosophy does not demand that we overcome cancer or heart disease through free will, but we must overcome behavioral disorders through mentalistic means. The end result of the criminal law/criminal justice system is punishment, prisons, and executions.

The utilitarians Bentham, Beccaria, and others argued that the basis for punishment

161

must be some future good to society such as the deterrence of criminal activity and the control of the crime rate (Jenkins, 1984; Mannheim, 1972; Rennie, 1978). Free will and moral responsibility were made part of the deterrence argument. The British legal philosopher H. L. A. Hart (1968:231) wrote that "a person may be punished if, and only if, he has voluntarily done something morally wrong; secondly, his punishment must match, or be the equivalent of, the wickedness of his offense; thirdly, the justification for punishing men under such conditions is that the return of suffering for moral evil voluntarily done, is itself just or morally good." Punishment as a means of deterring criminal behavior has failed miserably (Gibbs, 1975, 1979; Tittle, 1980). Blumstein, Cohen, and Nagin (1978: 7) concluded from their review of deterrence research that "in summary, therefore, we cannot yet assert that the evidence warrants an affirmative conclusion regarding deterrence."

The scientific model of human behavior is based on determinism, treatment, and prevention. Human behavior is the result of complex interaction among genetic and environmental variables. Out of the scientific model, or positive school of criminology, came biological, psychological, and social explanations of criminal behavior. The early biological theories of Cesare Lombroso and geneticists were never pursued either as a central point in criminological theory or as an important aspect of public policy for controlling crime. The scientific school of criminology led to the juvenile-court movement, the indeterminate sentence, the increased use of probation, and the use of probation in place of imprisonment. The prison system came to be the major institution for the control of crime and criminals. The purpose of criminal law was to rehabilitate offenders, not to punish them. Science must be used to control criminal behavior, not punishment and prisons.

Two major attempts at rehabilitating criminals occurred. One effort was related to psychiatry and the use of clinics and therapeutic programs to rehabilitate delinquents and criminals. The other was based on social theories of crime that stated that crime was a product of poverty, unemployment, and lower-class status. The effort to rehabilitate criminals did not work, and as a result of the failure of the rehabilitative ideal there was a return in the 1970s and 1980s to a retributive "law-and-order" model of criminal justice (Allen, 1964; Martin, Sechrest, and Redner, 1981). Lipton, Martinson, and Wilks (1975) published a report on the effectiveness of rehabilitative programs, and in this report they concluded that "nothing works." In response to the "failure of the rehabilitative ideal," several leading legal scholars and political scientists defended the use of prisons and punishment for the control of crime (Morris, 1974; Wilson, 1975). Wilson ended his book with the statement, "Wicked people exist. Nothing avails except to set them apart from innocent people" (1975: 209). The law-and-order model of crime control emerged in its present form in the 1964 Barry Goldwater presidential campaign and was the basis for the Law Enforcement Assistance Administration (LEAA) under Richard Nixon's administration. Under the LEAA program millions of federal dollars were poured into state projects to strengthen the police-courts-corrections systems as we returned to relying on prisons and executions to control the crime problem.

Research into the causes of criminal behavior was abandoned in favor of the "just-retribution" model of justice. Probation, parole, indeterminate sentences, and judicial discretion were abandoned or severely limited. Prisons became seriously over-crowded, and as a result most prison systems today are under court orders to correct problems of overcrowding, violence, violations of basic constitutional rights, and a lack of basic medical care. Today most states are using temporary facilities to house inmates, and in the State of Florida inmates must be released, as part of an early release plan, before other inmates are admitted to the prison system. The state also uses house arrest and electronic surveillance to keep inmates out of the prison system.

The failure of the LEAA program resulted in its abolition by Jimmy Carter's administration in 1978. Although the LEAA program was regarded as a failure (Cronin, Cronin, and Milakovitch, 1981; Feeley and Sarat, 1980; White and Krislov, 1977), we are still pursuing a "get tough on crime" position. The 1988 campaign between George Bush and Michael Dukakis was one of seeing which candidate could take the strongest law-and-order approach to the crime problem, each promising to support the police, get tough on drug suppliers, and make more use of more prisons. The campaign manager for the Dukakis campaign, Susan Estrich, has called for a tougher position on the crime problem (*Newsweek*, June 27, 1988; Estrich, 1987).

THE NEW SCIENCES OF BEHAVIOR

In the past, criminology has been primarily sociology, but in recent years there has been a major attempt to develop an interdisciplinary criminology based on biology, psychology, sociology, and law (Jeffery, 1990; Wilson and Herrnstein, 1985). The new sciences of behavior include behavioral genetics, the brain sciences, and learning theory. These new sciences of behavior must be integrated into an interdisciplinary system with the environmental sciences of geography, urban-planning ecology, and sociology and with the policy sciences of law, ethics, and political science.

The genetic basis for violent antisocial behavior is by now well established (Mednick and Christiansen, 1977; Mednick, Moffitt, and Stack, 1987; Plomin, DeFries, and McClearn, 1980; Singer, 1985). The genetic basis for male-limited alcoholism has also been established (Galanter, 1985; Goodwin, 1988; National Institute on Alcohol Abuse and Alcoholism, 1985).

The role of the brain in behavior is critical since the brain receives, codes, stores, and utilizes information from the environment in order to control behavior. I will review several of the major brain functions that are involved in the explanation of criminal behavior.

The peripheral nervous system is divided into somatic and autonomic. The somatic nervous system controls the large striated skeletal muscles, such as those in the arms and legs. The autonomic nervous system controls the smooth muscles of the internal organs, such as the heart, kidneys, blood vessels, liver, lungs, and so forth (Hoyenga and Hoyenga, 1988:20). We usually refer to the somatic nervous system as the "voluntary" system; that is, individuals voluntarily move their arms or legs, or their "behavior" is voluntary. What we refer to as "behavior" is the motor and sensory activities of the somatic nervous system. The autonomic nervous system is referred to as "involuntary"; it is beyond the control of the individual (e.g., blood pressure, heart beat, respiration rate, digestion). We do not usually refer to heart beat or blood pressure as "behavior." The law makes a critical distinction between a person taking a gun into his or her hand and firing it at another person and a hand jumping during an epileptic seizure, though both events are behaviors and are a product of the brain. What we call "free will" is the activities of the brain in controlling the somatic nervous system.

The "flight-or-fight" response is controlled by the autonomic nervous system. The limbic system is a group of interrelated parts of the forebrain that control the emotional aspects of behavior: anger, fear, hatred, love, sex, violence, and happiness. The pain and pleasure centers of the brain are located in the limbic system and related areas of the brain. These areas are responsible for the emotional and motivational aspects of our behavior (Carlson, 1986; Hoyenga and Hoyenga, 1988; Kalat, 1988).

An underaroused autonomic nervous system has been implicated in antisocial behavior. Learning the responses that avoid punishment, that is, learning law-abiding behavior, depends on the arousal of the autonomic nervous system, which in turn creates anxiety and avoidance behaviors. It has been discovered that antisocial individuals are often characterized by an underaroused autonomic nervous system (Mednick et al., 1982; Venables, 1987).

Brain laterality is also critical to antisocial behavior. The brain is divided into left and right hemispheres, with the left hemisphere controlling language and logical behaviors, the right hemisphere controlling mathematical and spatial functions. The two hemispheres are joined and normally share information; however, in some instances there may be hemispheric dysfunction. Most individuals (90%) are right-handed and left-hemisphere dominant. However, there are individuals with right-hemisphere dominance who are left-handed. Males are more right-hemisphere dominant than females because testosterone slows the development of the left hemisphere during fetal development (Kalat, 1988; Kolb and Whishaw, 1985). Some researchers have reported a higher incidence of left-handedness among criminals, whereas others have reported left-eye preference among criminals (Nachson and Denno, 1987).

Perhaps more critical than left- or right-hemisphere dominance is left-hemisphere dysfunction in the frontal and temporal lobes. The frontal and temporal lobes control auditory and visual sensory processing, long-term memory, emotional behavior, and long-range planning. The integration of emotional behavior with learning and memory and planning occurs in the temporal and frontal lobes (Kolb and Whishaw, 1985). Yeudall et al. (1982) found that 76% of the violent offenders and 91% of the psychopaths (Yeudall et al., 1987) had left-hemispheric dysfunction in the temporal and frontal lobes.

A very high rate of injuries and brain trauma has been found among criminals and delinquents (Lewis, 1981; Yeudall et al., 1985). These injuries can be the result of child abuse, the violent world of the delinquent, and/or a lack of adequate medical care. Violent offenders have four times the number of abnormal EEG (electroencephalogram) readings as offenders with only one offense. Yeudall found that 70% of the violent offenders he examined had suffered from head injuries (Mednick et al., 1982; Volavka, 1987). Lewis (1988), from an examination of 14 juveniles who were on death row, found that all 14 had brain trauma and neurological disorders. Nine had major neurological impairments, 7 had serious psychiatric disorders, 12 had IQ scores below 90, 12 had been brutalized sexually and physically as children, and 13 came from families with a history of violence. One of the most remarkable aspects of this study was the discovery that in only five cases did the lawyers have any kind of pretrial evaluation, and in those cases the lawyers ignored or hid the neurological and psychiatric evidence.

The brain communicates by means of the neurotransmitter system, or the "liquid nervous system." The major neurotransmitters are acetylcholine (ACh), norepinephrine (NE), epinephrine, and serotonin (5-HT). ACh increases affective (emotional) aggression, as well as predatory aggression. Dopamine and norepinephrine increase emotional and sexual aggression. As described by several contributors to this volume, serotonin decreases all types of aggression (see also Cloninger, 1987; Moyer, 1987; Rubin, 1987; Valzelli, 1981). It may be possible to control violent behavior by increasing the level of serotonin in the brain. Some of the most exciting work in neurology today is in relation to human behavior and the role of the neurotransmitters, alcohol, and drugs.

Diet and pollution are also critical to brain functioning. The precursors to the neurotransmitters are products of the diet: choline for acetylcholine; tyrosine for dopamine, norepinephrine, and epinephrine; and tryptophan for serotonin. What we eat is converted into the chemicals of the brain (Kalat, 1988). An excess of refined carbohydrates in the diet can result in a condition known as hypoglycemia, or low blood sugar. Since the brain is a major consumer of sugar, a lack of sugar in the brain can result in major behavioral difficulties, including confusion, irritability, irrationality, anxiety, and depression, as well as violence (Fishbein and Pease, 1988). Hypoglycemia is also associated with alcoholism and premenstrual syndrome. It is also known that alcohol reduces serotonin levels in the brain. Carbohydrates increase while protein decreases the levels of tryptophan reaching the brain to be converted into serotonin (Fishbein and Pease, 1988).

Trace elements also impact the brain and thus behavior. The good trace elements are zinc, calcium, magnesium, selenium, iron, and potassium; the bad are lead, cadmium, arsenic, aluminum, and mercury. The bad trace elements can impact the growth and development of the brain. Hyperactivity and low IQ are the result of lead and cadmium concentrations in the brain (Fishbein and Pease, 1988; National Academy of Sciences, 1980). Hair analyses of serial murderers and violent offenders have shown excessive concentrations of lead and cadmium in such individuals, including Henry Lee Lucas and James O. Huberty, the man who killed 21 persons at a McDonald's restaurant in San Diego (Pihl et al., 1982; Walsh, 1984, 1985).

Sociologists are inclined to discuss criminality in terms of poverty and unemployment. One of the most obvious links is between poverty and diet, pollution, and the brain. Poor people suffer more brain damage from diet and pollution than do rich people.

SEX DIFFERENCES AND THE BRAIN

At conception males and females are differentiated genetically; the typical male is an XY, whereas the typical female is an XX. The male chromosome sends out testosterone to the fetus, which then results in a different body and a different brain. The male is much more susceptible to fetal abnormalities, higher rates of disease, higher rates of death, and higher rates of behavioral disorders, such as suicide, mental illness, and criminal behavior.

Testosterone produces a male brain that is often right-hemisphere dominant, as was noted above. Brain development depends on the level of testosterone produced by the mother at the time the fetus is developing. High levels of testosterone during the fetal development of a female will result in an androgenized female, or a female with the physical and behavioral characteristics of the male, or a masculinized female. The same thing occurs after birth if the female has excess testosterone because of the malfunctioning of her adrenal glands (Carlson, 1986; Kalat, 1988; Money and Ehrhardt, 1972; Parsons, 1980). A condition called androgen insensitivity occurs in the male if during the fetal stage the masculine brain does not develop the proper receptor sites for testosterone. Regardless of testosterone levels in the blood, these individuals will not respond to testosterone in a normal manner. The question is often raised as to whether or not homosexuals have normal circulating testosterone in the blood. What such studies ignore is that a normal testosterone level in the blood stream does not mean normal testosterone level in the brain. Testosterone levels have been related to violence, and the male is much more violent than is the female. In the case of premenstrual syndrome, where the female is sometimes violent, there is a disturbance in the balance of the sex hormones.

THE VIOLENT YOUTHFUL OFFENDER

The Violent Offender Project at Ohio State University (Conrad and Dinitz, 1977) found that 61% of the violent offenses were committed by chronic offenders who started their careers at an early age, usually before 13. These offenders often had five offenses or more (Hamparian et al., 1978). The project concluded that, if violent offenders were given a flat-time sentence as advocated by the just-retribution school (see above discussion), violent crime would be reduced overall by 25%, whereas the prison population would increase by 500% (Van Dine et al., 1979).

Recent criminological studies have focused on the repeat offender who commits most of the serious crimes. Wolfgang et al. (1972) found that 6% of the offenders committed 50% of the offenses. The same conclusion has been reached by Robins (1966), McCord and McCord (1959), West and Farrington (1973, 1977), and Glueck and Glueck (1950, 1968). Farrington, Ohlin, and Wilson (1986) summarize this work, which is longitudinal in design and uses incidence data rather than prevalence data; that is, the focus is on the development of individual careers over a long period of time.

Recent research has focused on the violent career offender (Blumstein et al., 1986; Chaiken and Chaiken, 1982; Petersilia et al., 1977) These studies have consistently found that there is a small group of offenders who commit a majority of the crimes. These individuals start their careers before age 12; they have committed five or more offenses by age 18; they do not stop offending as do most offenders at age 18; and they are involved in alcohol and drug use. Social variables used by sociologists, such as sex, age, race, and socioeconomic status, are only weakly associated with the violent career offender. This indicates that possibly the career offender is different in kind (Blumstein et al., 1986; Chaiken and Chaiken, 1982). These individuals may differ in terms of the genetic and neurological factors discussed above.

A major study of the violent career offender must be started, using an interdisciplinary team of geneticists, neurologists, psychiatrists, psychologists, and lawyers. We must learn to identify such individuals early in their careers so that they can be put into treatment programs. The usual legal response to the career criminal is legislation creating a special prosecutor and long prison sentences for those who are defined as career criminals. Such action does not prevent crimes; it only results in more work for the police, more work for the prosecutors, and more prisons for the criminals.

PUNISHMENT AND PREVENTION

Two major issues emerge from the conflict between science and law. One issue is that of punishment versus treatment and prevention; the other is the insanity defense.

Criminal law is based on a primitive concept of revenge. Early tribal law was founded on the blood feud wherein justice was in the hands of the kinship group. Unregulated violence was the result of the blood feud, and in order to control the feud a compulsory system of compensation similar to our modern tort law emerged. During the 10th and 11th centuries, when the state emerged to replace the tribal system that by then had decayed, the king made and enforced the law of the land. At this time tort law came to be distinguished from criminal law. Tort law involved a private wrong that was resolved by compensation. Criminal law involved a public wrong against society for which punishment and not compensation was required. Criminal law picked up the banner of revenge in the name of the state. Individuals could no longer seek revenge for harms done against them; victims of crimes must depend on the state to arrest, prosecute, convict, imprison, and execute criminals.

Criminal law did not end violence; it only transferred it from the private sphere to the public sphere. I would make a speculative judgment that, in the evolution of the primate brain, the rational brain evolved along side the emotional brain. As the rational brain grew in size and complexity it gained some control over the emotional brain. However, the rational brain never did gain complete control over the emotional brain; it only placed violence within the confines of rational conceptualizations, which justified punishment and violence in terms of such basic legal doctrines as mens rea, moral guilt, just retribution, and utilitarian deterrence. We still execute criminals, but we do so with great legal ceremony and dialogue. The brutality of the old revenge is retained, but it is hidden from public view. Our prison systems and executions are as brutal as medieval punitive systems, but they are given respectability by the rational brain.

We know that chimpanzees will kill other chimpanzees, but there is no guilt or group revenge or group hatred against the killer. There is no memory, nor are there feelings of an injustice that must be revenged for Kantian ethics to prevail (Goodall,

1986). I do not know what goes on in the chimpanzee's brain; but the human brain carries around feelings of hatred and revenge towards the trespasser or killer. Whatever differences exist between the human brain and the chimpanzee brain, and they are primarily concerned with language, logic, ethics, and other higher-order thought processes, these differences have created our major legal, religious, and philosophical systems. Future research in law and biology should include this problem as an important part of its agenda.

If one accepts the view that criminal law is a product of the evolution of the rational brain, then one can consider the next evolutionary stage, which is that the rational brain creates a system of law and justice that is not based on violence. My argument is based on the observation that our present criminal law/criminal justice system is motivated by, and based on, emotions (e.g., those emotions that are involved in the hatred of the criminal). There is nothing rational about putting individuals in prison or executing them for their behavior. Such a system of justice is costly and ineffective. It does not deter or rehabilitate, nor does it protect the public, the victim, or the criminal.

When I say the system is not rational I mean the system does not achieve rational goals, such as control of crime or the establishment of social order. If we wanted to turn violent men into law-abiding citizens, we would not put them in prisons; if we wanted to create violent men we would put them in prisons.

The present criminal justice system is not ethically defensible, even though it is put forth as a system derived from ethical principles. We need a new examination of the ethics of criminal law, one based on the scientific principles of behavior. We approach the philosophy of law from the standpoint of a dualism of mind and matter. Natural-law philosophers do not derive law from science or from behavior. What "is" is separate from "what ought to be." The legal positivists found the law in the behavior of kings. The legal realists attempted to find the law in the behavior of individuals, which could be understood if law interacted with science. The sociological school, led by Roscoe Pound, looked at law as a means of social control and social engineering. Law-school education is sadly lacking in interdisciplinary materials related to law and the behavioral sciences. We must work to establish interactions among lawyers, philosophers, geneticists, neurologists, psychologists, criminologists, and other behavioral scientists. There is at present an attempt to join philosophy, ethics, biology, and neurology in such works as *The Biology of Moral Systems* (Alexander, 1987); *Mind and Brain* (LeDoux and Hirst, 1986); and *Neurophilosophy* (Churchland, 1986).

Another ethical issue that must be addressed is the concept of responsibility for one's actions. Individuals are responsible for their actions to the extent that they are able to evaluate and make decisions about the consequences of their behavior. Responsibility means that the rational brain is in control of the emotional brain. A brain that lacks certain biochemicals or certain past environmental experiences is not capable of making rational decisions. Since behavior is dependent on the serotonin levels of the brain, whether or not one is rational depends on the serotonin levels of the brain. I may be calm or irrational depending on the serotonin level. My moral self is thus defined by my serotonin level.

CRIME PREVENTION AS AN ALTERNATIVE CRIME-CONTROL MODEL

Crime prevention has been ignored in criminology and in criminal justice; however, recently there has been a renewed effort to develop crime-prevention programs (Brantingham and Brantingham, 1981, 1984; Clarke and Mayhew, 1980; Harries, 1980; Jeffery, 1971,1977; National Crime Prevention Institute, 1986; Poyner, 1983). This effort has taken the form of redesigning streets, parks, buildings, schools, residential areas, and business areas. A simple but dramatic example is the reduction of convenience store robberies by 64% in Gainesville, Florida, by requiring two clerks to be present in the store in place of the usual one (Clifton, 1987).

Crime prevention must involve the individual offender, as well as the physical environment. As I stated above, we must establish a research project for the early identification of individuals who are destined to be youthful violent offenders. Such individuals can be identified by the age of 7 or 8. Unless we establish crime-prevention programs, we will continue to have more crime, to arrest more criminals, to build more prisons, and to have more victims of crime.

LEGAL BARRIERS TO CRIME-PREVENTION PROJECTS

As I have noted, the criminal justice system is based on revenge and punishment. We have legalized the "right-to-punishment" doctrine without any attempt to create a "right-to-treatment" doctrine. The U.S. Constitution provides certain protections for the accused before he or she can be punished: no cruel and unusual punishment, the right to a speedy trial, the right to a jury trial, the right to hear the evidence against oneself, the right to avoid confessions and testimony against oneself, and so forth. The state must follow procedural criminal law before imprisoning or executing a citizen, and for capital-offense cases we have an elaborate appeals process, which takes years and costs millions of dollars to achieve.

We have, however, developed a right-to-treatment doctrine for the mentally ill (Birnbaum, 1982; Brooks, 1974; Jeffery, 1985; Miller et al., 1976). There is also a right-to-refuse-treatment doctrine, which means that the patient may refuse psychotropic drugs, electroconvulsive therapies, behavior-modification therapies, and psychosurgery (Brooks and Winick, 1987; Shapiro, 1982; Shapiro and Spice, 1984). If the person has a right to refuse treatment for mental diseases, then that person must also possess a right to refuse treatment for criminal behavior. Compulsory treatment of criminals is opposed by many legal scholars (Morris, 1974; Shapiro, 1982; Shapiro and Spece, 1984; Wilson, 1975). We are not allowed to do research on violent offenders or to use therapies to treat them for their behavior (Gaylin, Macklin, and Powledge, 1981). We cannot stick a needle into the arm of a criminal to determine if he or she is hypoglycemic, but we can to execute that criminal. Judge David Baselon stated in *United States v. Alexander*, 471 F.2d. 923 (1973) that we allow only those therapies that are guaranteed to fail; we do not allow those therapies that might work.

The major objections are to the "organic therapies," those therapies based on biological rather than mentalistic premises. The legal arguments for rejecting organic therapies include the First Amendment and the freedom of speech, which has been interpreted to mean the right to mentation and the ability to generate one's own thoughts; the right to privacy; the right to be free from coercion; and the right to informed consent. The autonomy of the individual is protected even if it means allowing that individual to remain a dangerous and violent person (Shapiro, 1982). Shapiro notes, for example, that the issue is "what do I wish to be" (1982: 26); or "forcing people to do what they do not wish to do is prima facie wrong" (1982: 53).

Several points can be made concerning the rejection of physical or medical treatment for criminals. The law that protects the rights of the individual to freedom of speech and mentation, to privacy, to freedom from coercion, and to informed consent does not protect the criminal. The freedom of mentation is not protected for those who are placed in prisons or in electric chairs. They are forced to do what they do not want to do. It is impossible to say what free will or thoughts are protected by prisons and electric chairs. It is hard to determine how much brain tissue is destroyed by imprisonment; we do know how much is destroyed by electrocution. Criminals have no right to informed consent. Criminals must consent to treatment, but they do not and cannot consent to punishment. Another problem with the legal arguments is that they assume that individuals are free to be what they want to be. But we are determined by our genetic and environmental backgrounds. That I want to be a great basketball player or concert pianist or chemist has nothing to do with my abilities to achieve such goals. The right to mentation is a meaningless right for someone with a defective brain (e.g., with a biochemical imbalance or a damaged brain or a defective neurotransmitter system). The goal of the organic therapies is to make well the sick brain. The person who has a frontal or temporal lobe disorder is unable to decide "freely" how he or she is going to behave. The purpose of organic therapies is to make the individual competent to make decisions on his or her own, decisions that cannot be made with a defective brain. A standard of treatment that could be used is one that states that the treatment must do more good than harm. This is a standard that if applied to our criminal justice system would effectively block any further use of prisons for the solution of our crime problem.

If we are to create a treatment system for criminals in place of the punishment system, we must devise a legal justification for compulsory treatment. We have compulsory punishment for criminals, which violates the free will and mentation of the criminal, but we do not allow compulsory treatment. If criminals do not want to be treated, then we must either release them untreated or put them in prison. In either case we have not treated the criminal or prevented future crimes from occurring. We can (1) require medical treatment for criminals, (2) give criminals a choice of imprisonment or medical treatment, or (3) punish criminals without the possibility of treatment or prevention.

The criminal law requires medical treatment of criminals. A major decision in this area is *Ruiz v. Estelle*, 503 F. Supp. 1265 (1980). The concept of medical treatment extends to heart disease and cancer, but it has never been extended to the right to have a CAT scan or a PET scan or an examination for hypoglycemia. Criminals do not have the right to a special diet if they are hypoglycemic. The right to medical treatment must be extended to the field of neurology and newer concepts of a brain-behavior relationship.

A related topic is that of informed consent (Brooks and Winick, 1987; Horan and Milligan, 1983; Valenstein, 1980). The doctrine of informed consent applies to the area of consent by a patient to private medical treatment. We follow J.S. Mill's standard of "harm to others" in deciding where to interfere with human liberty. A person is allowed to do harm to self if such behavior is not harmful to others. We allow a person to smoke, but not in an area where it will cause harm to others. Today a critical issue is the question of whether or not the victims of AIDS can be forced into compulsory treatment and/or confinement and isolation from the general public.

BIOLOGICAL PSYCHIATRY AND THE INSANITY DEFENSE

The insanity defense is another area where science and law are in conflict. Insanity is a legal concept and a legal defense based on the inability of the defendant to form the necessary criminal intent or mens rea at the time of the crime.

The basic rule is found in the M'Naghten case (1843), which is known as the "right and wrong" rule, or the ability of the defendant to distinguish right from wrong. This constitutes a *rational* test of mens rea; that is, it is a test of the ability of the brain to function rationally (Bromberg, 1979; Brooks, 1974; Brooks and Winick, 1987; Robinson, 1980).

The other test for insanity is irresistible impulse, or the inability of the defendant to control his or her behavior at the time of the crime because of an irresistible impulse. This is an *emotional* test for insanity. This test goes to the limbic system or the emotional aspect of the brain (Bromberg, 1979; Jeffery, 1985). The American Law Institute (ALI) Model Penal Code combines the rational test and the emotional test by stating that "a person is not responsible for criminal conduct if at the time of such conduct as a result of mental disease or defect he lacks substantial capacity either to appreciate the criminality of his conduct or to conform his conduct to the requirements of the law" (Low, Jeffries, and Bonnie, 1982, appendix A). The Model Penal Code notes that only voluntary acts are subject to penal sanctions. Involuntary acts, such as reflexive acts or convulsions, are not regarded as voluntary acts.

The concept of insanity is not scientifically defensible or testable. The usual procedure is the "battle of the experts," or the procedure whereby psychiatrists for the government testify that the defendant knows right from wrong, and psychiatrists for the defense testify that the defendant did not know right from wrong. The jury must then decide what constitutes insanity. The criminal trial utilizes psychiatrists not as experts on brain damage but as combatants in the adversarial system.

The concept of mental illness is used by psychoanalytically oriented psychiatrists. They observe behavior, either in terms of psychiatric interviews or in terms of subjective tests, such as the TAT or Rorschach. Abnormal behaviors are then labeled schizophrenic or manic-depressive. Such diagnostic categories are not accurate or scientifically valid (Gaylin, 1982; Jeffery, 1985; Roche, 1958; Ziskin, 1975; Ziskin and Faust, 1988). If the defendant is found guilty, he or she is sent to prison, possibly to return to the streets more dangerous than ever. If the defendant is found not guilty by reason of insanity, he or she is placed in a mental hospital, which may be worse than the prison, and where he or she may receive little or no treatment. One of the major arguments against the insanity defense is that it allows dangerous people to be released onto the streets without adequate treatment, whereby they commit more violent crimes. Labeling people legally insane does not mean they will be treated successfully. On the other hand, placing individuals in prison does not guarantee they will be given treatment and will be cured when they are released.

Psychology and psychiatry have shifted in recent years from a mentalistic to a physicalistic model based on the new brain sciences. Freud predicted in his *Project for Scientific Psychology* that psychiatry would eventually rejoin medicine and be a part of neurology (Pribram and Gill, 1976; Rychlak, 1981; Sulloway, 1979). Today CAT scans and NMR scans and drug therapies are used in the diagnosis and treatment of behavioral disorders. The concept of a sick brain replaces that of the sick mind (Andreasen, 1985; Curzon, 1980; Jacobs and Gelperin, 1981; Lickey and Gordon, 1983; Meltzer, 1979; Reinis and Goldman, 1983; Snyder, 1980; Wender and Klein, 1981).

In recent years several neurological conditions have been introduced into criminal trials as part of the insanity defense, for example, epilepsy, premenstrual syndrome, hypoglycemia, post-traumatic stress syndrome, and psychopathy (Jeffery, 1985). An unusual case involving the impact of alcohol on the brain is told in *The Crocodile Man* (Mayer and Wheeler, 1982). A young college student, male, attacked and almost killed two women in a bar one evening after ingesting a small amount of beer. The father of the defendant, who happened to be a prominent biochemist at a northeastern university, felt that there was a biochemical defect in his son's brain that caused this sensitivity to alcohol. After months of research, a defect in the liver enzyme system that processes alcohol was found.

One of the most interesting uses of neurological testimony in a criminal trial occurred in the trial of John Hinckley, Jr., the man who shot President Ronald Reagan in 1981. The government tried to show that Hinckley was rational (that he knew right from wrong) at the time of the crime because he had written to actress Jodie Foster, he had purchased a gun, and he had tracked the president for several months. Government psychiatrists also testified that Hinckley had a mood disorder (Low, Jeffries, and Bonnie, 1986).

The defense psychiatrists testified that Hinckley was schizophrenic or perhaps a borderline personality (Low, Jeffries, and Bonnie, 1986). The most interesting aspect of the case, and one totally ignored by the Low, Jeffries, and Bonnie discussion of the case, was the testimony of a young biological psychiatrist from Harvard University Medical School by the name of David Bear (Caplan, 1984). Bear attempted to introduce into the trial a CAT scan taken of Hinckley's brain at Duke University Medical Center where Hinckley had been sent for a psychiatric evaluation. The CAT scan revealed that the ventricles of Hinckley's brain were abnormally enlarged with major portions of the brain missing. The judge ruled that the CAT scan could not be admissible as evidence. At that point a major argument occurred between Bear and the judge, and Bear refused to appear as an expert witness unless he was allowed to introduce into the trial the scientific evidence he used in coming to his conclusions about the status of Hinckley's brain. The judge then allowed the CAT scan to be introduced as evidence. I do not know why the jury brought in a decision of not guilty by reason of insanity, but I have to believe it was because the CAT scan showed a major brain defect, in contrast to the psychobabble of the other psychiatrists about mood disorders and borderline personalities. The jury elected to follow the lead of modern biological psychiatry.

Two observations can be made about the use of the insanity defense by lawyers. One of the peculiar results of the insanity defense and modern biological psychiatry is that such physical defects of the brain as epilepsy and premenstrual syndrome are being called evidence of insanity rather than what they really are: neurological defects. Another unfortunate aspect of the insanity defense turns up when neurological defects are introduced into a criminal trial but the verdict is *not* not guilty by reason of insanity; rather, the medical condition is introduced as evidence of diminished responsibility, and the defendant is given a reduced sentence: seven years, rather than 15, in a prison because of brain damage. The defendant is given less time in prison and is then released back into the community with his or her damaged brain (Jeffery, 1985).

Finally, it must be observed that the law has in its own curious way followed several of the basic principles of brain organization. As I have noted, the brain is divided into a rational brain and emotional brain, a conceptual division that the insanity defense follows in its references to rationality and emotionality as bases for the insanity defense. Also, I noted that the peripheral nervous system is divided into somatic and autonomic systems. The somatic system, concerned with the large striated skeletal muscles, we often refer to as a " voluntary system." On the other hand, the autonomic system, which controls our heart rate, blood pressure, and digestive processes, is called "involuntary." The law does not call our heartbeat, or our reflexive acts, voluntary. The law uses the division of the nervous system into somatic and autonomic as a basis for talking about voluntary versus involuntary behaviors.

LAW AND BIOLOGY: THE FUTURE

There is a great need for future interdisciplinary efforts to integrate law and biology. This can be done through conferences, such as the one that led to this volume, or through seminars, such as the one at Yale University Law School under the direction of Professor E. Donald Elliott.

There is a need for a large research project, preferably associated with a leading law school and a major medical research institute or medical center, devoted to the diagnosis and treatment of violent youthful offenders. Such a research effort must involve the application of knowledge from the brain sciences to the control of criminal behavior. There is an institution in Florida for special juvenile violent offenders with long histories of violent crimes and substance abuse, but it is at present doing nothing for these youths. The same thing can be said for most other state penal systems.

Such a research effort must be devoted not only to biological issues but also to the legal and ethical issues that are involved in a treatment model. A right-to-treatment model as discussed by Kittrie (1971) must be developed before we can make use of the brain sciences within our criminal justice system. We must shift the emphasis from punishment to treatment and prevention. Crime prevention must replace the police-courts-prison system. Basic changes in the legal and political systems are required, however, before this will be accomplished. I hope this book will serve as a vehicle for such changes.

A RESPONSE TO PROF. MICHAEL H. SHAPIRO

In his excellent chapter, Professor Shapiro has grasped and presented the legal side of the argument as involved in the conflict between law and science. Without going into a detailed critique of his chapter, I would like to make a basic observation concerning the nature of his legal argument.

He starts by stating that the brain sciences will have no impact on legal ideas of responsibility, free will, and freedom. This argument is based on the philosophical position of mind-body dualism, which I discuss in detail in this chapter. Shapiro notes that neurological impairments will not count under an insanity or diminished-responsibility defense unless such impairments impact on the moral and legal level. He argues that the physical substrate must in some way connect to legal notions of right and wrong and mens rea.

To state, as Shapiro does, that the impairment must be translated into insanity or diminished responsibility is to ignore that such concepts as insanity, free will, and freedom are creations of the mind and not of the brain. If behavior is a product of the brain, then it is physically impossible for criminal behavior not to be controlled or influenced in some way by neurochemical malfunctioning. The argument that ideas of freedom, free will, and responsibility are neutral in respect to the modern brain sciences is only tenable if we hold that behavior is controlled by the mind and not the brain. I argue that behavior is controlled by the brain, even those behaviors with which lawyers deal, and we must reinterpret such legal concepts as insanity and free will into physical, neurological concepts. Medical definitions must replace legal definitions.

REFERENCES

Alexander, R. (1987). *The Biology of Moral Systems.* Hawthorne, NY: Aldine de Gruyter.

Allen, F.A. (1964). *Borderline of Criminal Justice.* Chicago: University of Chicago Press.

Andreasen, N. (1985). *Broken Brain.* New York: Harper and Row.

Birnbaum, M. (1982). "The Right to Treatment." In *Critical Issues in American Psychiatry and the Law*, ed. R. Rosner. Springfield: Thomas.

Blumstein, A., et al. (1986). *Criminal Careers and Career Criminals.* Vol. 1. Washington, DC: National Academy Press.

Blumstein, Cohen, and Nagin. (1978). *Deterrence and Incapacitation: Estimating the Effects of Criminal Sanctions on Crime Rates.* Washington, DC: National Academy of Sciences.

Brantingham, P.J., and P.L. Brantingham. (1981). *Environmental Criminology*. Beverly Hills: Sage.

Brantingham, P.J., and P.L. Brantingham. (1984). *Patterns in Crime*. New York: Macmillan.

Bromberg, W. (1979). *The Use of Psychiatry in the Law*. Westport: Quorum Books.

Brooks, A.D. (1974). *Law Psychiatry and the Mental Health System*. Boston: Little Brown.

Brooks, A.D., and B.J. Winick (Eds.). (1987). "Current Issues in Mental Disability Law." *Rutgers Law Review* 39:1–244.

Caplan, L. (1984). *The Insanity Defense and the Trial of John W. Hinckley, Jr.* Boston: Godine.

Carlson, N.R. (1986). *Physiology of Behavior*. Boston: Allyn and Bacon.

Chaiken, J.M., and M.R. Chaiken. (1982). *Varieties of Criminal Behavior*. Santa Monica: Rand.

Churchland, P.S. (1986). *Neurophilosophy: Toward a Unified Science of the Mind-Brain*. Cambridge: MIT Press.

Clarke, R.V.G., and P. Mayhew. (1980). *Designing Out Crime*. London: Her Majesty's Stationery Office.

Clifton, W., Jr. (1987). *Convenience Store Robberies in Gainesville, Florida*. Gainesville: Gainesville Police Department.

Cloninger, R.D. (1987). "Pharmacological Approaches to the Treatment of Antisocial Behavior." In *The Causes of Crime: New Biological Approaches*, ed. S. Mednick, T. Moffitt, and S.A. Stack. Cambridge: Cambridge University Press.

Conrad, J., and S. Dinitz. (1977). *In Fear of Each Other*. Lexington: Lexington Books.

Cronin, T.E., T.Z. Cronin, and M.E. Milakovitch. (1981). *U.S. versus Crime in the Streets*. Bloomington: Indiana University Press.

Curzon, G. (Ed.). (1980). *Biochemistry and Psychiatric Disturbances*. New York: Wiley and Sons.

Estrich, S. (1987). *Real Rape*. Cambridge: Harvard University Press.

Farrington, D.P., L. Ohlin, and J.Q. Wilson (Eds.). (1986). *Understanding and Controlling Crime*. New York: Springer-Verlag.

Feeley, M.M., and A.D. Sarat. (1980). *The Policy Dilemma*. Minneapolis: University of Minnesota Press.

Fishbein, D., and S. Pease. (1988). *The Effects of Diet on Behavior: Implication for Criminology and Corrections*. Boulder: National Institute of Corrections.

Galanter, M. (Ed.) (1985). *Recent Developments in Alcoholism*. Vols. 1 and 3. New York: Plenum.

Gaylin, W. (1982). *The Killing of Bonnie Garland*. New York: Simon and Schuster.

Gaylin, W., R. Macklin, and T.M. Powledge (Eds.). (1981). *Violence and the Politics of Research*. New York: Plenum.

Gibbs, J. (1975). *Crime, Punishment, and Deterrence*. New York: Elsevier.

Gibbs, J. (1979). "Assessing the Deterrence Doctrine." *American Behavioral Scientist* 22 (6, July/August): 653–677.

Glueck, S., and E.T. Glueck. (1950). *Unraveling Juvenile Delinquency*. Cambridge: Harvard University Press.

Glueck, S., and E.T. Glueck. (1968). *Delinquents and Nondelinquents in Perspective*. Cambridge: Harvard University Press.

Goodall, J. (1986). *The Chimpanzees of Gombe*. Cambridge: Harvard University Press.

Goodwin, D.W. (1988). *Is Alcoholism Hereditary?* Belmont: Wadsworth.

Hamparian, D.M., et al. (1978). *The Violent Few*. Lexington: Lexington Books.

Harries, K. (1980). *Crime and the Environment*. Springfield: Thomas.

Hart, H.L.A. (1968). *Punishment and Responsibility*. New York: Oxford University Press.

Horan, D.J., and R.J. Milligan. (1983). "Recent Developments in Psychiatric Malpractice." *Behavioral Sciences and the Law* 1:23–28.

Hoyenga, K.B., and K. Hoyenga. (1988). *Psychobiology*. Pacific Grove: Brooks/Cole.

Jacobs, B.L., and P. Gelperin. (1981). *Serotonin, Neurotransmission, and Behavior*. Cambridge: MIT Press.

Jeffery, C.R. (1971; 1977). *Crime Prevention Through Environmental Design*. Beverly Hills: Sage.

Jeffery, C.R. (1985). *Attacks on the Insanity Defense: Biological Psychiatry and New Perspectives on Criminal Behavior*. Springfield: Thomas.

Jeffery, C.R. (1990). *Criminology: An Interdisciplinary Approach*. Englewood Cliffs: Prentice-Hall.

Jenkins, P. (1984). *Crime and Justice*. Belmont: Wadsworth.

Jones, D.A. (1986). *History of Criminology: A Philosophical Perspective*. New York: Greenwood.

Kalat, J.W. (1988). *Biological Psychology*. Belmont: Wadsworth.

Kittrie, N. (1971). *The Right to Be Different*. Baltimore: Johns Hopkins University Press.

Kolb, B., and I.Q. Whishaw. (1985). *Fundamentals of Human Neuropsychology*. New York: Freeman.

LeDoux, J.E., and W. Hirst. (1986). *Mind and Brain: Dialogues in Cognitive Neuroscience*. Cambridge: Cambridge University Press.

Lewis, D. (1981). *Vulnerabilities to Delinquency*. Jamaica, NY: SP Medical and Scientific Books.

Lewis, D., et al. (1988). "Neuropsychiatric, Psychoeducational, and Family Characteristics of 14 Juveniles Condemned to Death in the United States." *American Journal of Psychiatry* 145(5): 584–589.

Lickey, M.E., and B. Gordon. (1983). *Drugs for Mental Illness*. San Francisco: Freeman.

Lipton, D., R. Martinson, and J. Wilks. (1975). *Effectiveness of Correctional Treatment: A Survey of Treatment Evaluation Studies*. New York: Praeger.

Low, P.W., J.C. Jeffries, and R.J. Bonnie. (1982). *Criminal Law*. Mineola, NY: Foundation Press.

Low, P.W., J.C. Jeffries, and R.J. Bonnie. (1986). *The Trial of John W. Hinckley, Jr.* Mineola, NY: Foundation Press.

McCord, W.J., and J. McCord. (1959). *Origins of Crime*. New York: Columbia University Press.

Mannheim, H. (Ed.). (1972). *Pioneers in Criminology*. Montclair, NJ: Patterson-Smith.

Martin, S.E., L.B. Sechrest, and R. Redner. (1981). *New Directions in the Rehabilitation of Criminal Offenders*. Washington, DC: National Academy Press.

Mayer, A., and M. Wheeler. (1982). *The Crocodile Man*. Boston: Houghton-Mifflin.

Mednick, S., et al. (1982). "Biology and Violence." In *Criminal Violence*, ed. M. Wolfgang and N. Weiner. Beverly Hills: Sage.

Mednick, S., and H.O. Christiansen. (1977). *Biosocial Basis of Criminal Behavior*. New York: Gardner.

Mednick, S., T. Moffitt, and S.A. Stack. (1987). *The Causes of Crime: New Biological Approaches*. Cambridge: Cambridge University Press.

Meltzer, H.L. (1979). *The Chemistry of Human Behavior*. New York: Nelson Hall.

Michael, J., and M. Adler. (1933). *Crime, Law, and Social Science*. New York: Harcourt Brace.

Miller, F.W., et al. (1976). *The Mental Health Process*. Mineola, NY: Foundation Press.

Money, J., and A.A. Ehrhardt. (1972). *Man and Woman, Boy and Girl*. Baltimore: Johns Hopkins University Press.

Morris, N. (1974). *The Future of Imprisonment*. Chicago: University of Chicago Press.

Moyer, K.E. (1976). *The Psychobiology of Aggression.* New York: Harper and Row.

Moyer, K.E. (1987). *Violence and Agression.* New York: Paragon.

Nachson, S., and D. Denno. (1987). "Violent Behavior and Cerebral Hemisphere Function." In *The Causes of Crime: New Biological Approaches,* ed. S. Mednick, T. Moffitt, and S.A. Stack. Cambridge: University of Cambridge Press.

National Academy of Sciences. (1980). *Lead in the Human Environment.* Washington, DC: National Academy Press.

National Crime Prevention Institute. (1986). *Understanding Crime Prevention.* Stoneham: Butterworths.

National Institute of Alcohol Abuse and Alcoholism. (1985). *Alcoholism: An Inherited Disease.* Washington, DC: U.S. Government Printing Office.

Parsons, J.E. (1980). *The Psychobiology of Sex Differences.* Washington, DC: Hemisphere.

Petersilia, J., et al. (1977). *Criminal Careers of Habitual Felons.* Santa Monica: Rand.

Pihl, R.O., et al. (1982). "Hair Element Analysis of Violent Criminals." *Canadian Journal of Psychiatry* 7(6): 523–534.

Plomin, R., J.C. DeFries, and G.E.R. McClearn. (1980). *Behavioral Genetics.* San Francisco: Freeman.

Poyner, B. (1983). *Design Against Crime.* London: Butterworths.

Pribram, K.H., and M.M. Gill. (1976). *Freud's Project Reassessed.* New York: Basic Books.

Reinis, S., and J.M. Goldman. (1983). *The Chemistry of Behavior.* New York: Plenum.

Rennie, Y. (1978). *The Search for Criminal Man.* Boston: Lexington.

Robins, L.N. (1966). *Deviant Children Grown Up.* Baltimore: Williams and Wilkins.

Robinson, D.N. (1980). *Psychology and Law.* New York: Oxford University Press.

Roche, P.Q. (1958). *The Criminal Mind.* New York: Farrar, Straus, and Cudahy.

Rubin, R. (1987). "The Neuroendocrinology and Neurochemistry of Antisocial Behavior." In *The Causes of Crime: New Biological Approaches,* ed. S. Mednick, T. Moffitt, and S.A. Stack. Cambridge: University of Cambridge Press.

Rychlak, J.F. (1981). *Introduction to Personality and Psychotherapy.* Boston: Houghton-Mifflin.

Shapiro, N.H. (1982). *Biological and Behavioral Technologies and the Law.* New York: Praeger.

Shapiro, N.H., and R.G. Spece. (1984). *Bioethics and Law.* St. Paul: West.

Singer, S. (1985). *Human Genetics.* New York: Freeman.

Snyder, S.H. (1980). *Biological Aspects of Mental Disorders.* New York: Oxford University Press.

Sulloway, F. (1979). *Freud: Biologist of the Mind.* New York: Basic Books.

Tittle, C.R. (1980). *Sanctions and Social Deviance.* New York: Praeger.

Valenstein, E.S. (1980). *The Psychosurgery Debate.* San Francisco: Freeman.

Valzelli, L. (1981). *Psychobiology of Violence and Aggression.* New York: Raven.

Van Dine, S., et al. (1979). *Restraining the Wicked.* Boston: Lexington Books.

Venables, P. (1987). "Autonomic Nervous System Factors in Criminal Behavior." In *The Causes of Crime: New Biological Approaches,* ed. S. Mednick, T. Moffitt, and S.A. Stack. Cambridge: University of Cambridge Press.

Volavka, J. (1987). "Electroencephalogram Among Criminals." In *The Causes of Crime: New Biological Approaches,* ed. S. Mednick, T. Moffitt, and S.A. Stack. Cambridge: University of Cambridge Press.

Walsh, W.J. (1984). "Trace Metal Concentrations in the Hair of Henry Lee Lucas." Unpublished manuscript.

Walsh, W.J. (1985). "Elemental Concentrations in the Hair of James Oliver Huberty." Report to the Coroner, San Diego County, California.

Wender, P.W., and D.F. Klein. (1981). *Mind, Mood, and Medicine*. New York: Farrar, Straus, Giroux.

West, D.J., and D.P. Farrington. (1973). *Who Becomes Delinquent*. London: Heinemann.

West, D.J., and D.P. Farrington. (1977). *The Delinquent Way of Life*. London: Heinemann.

White, S.O., and S. Krislov. (1977). *Understanding Crime*. Washington, DC: National Academy of Sciences.

Wilson, J.Q. (1975). *Thinking About Crime*. New York: Basic Books.

Wilson, J.Q., and R. Herrnstein. (1985). *Crime and Human Nature*. New York: Simon and Schuster.

Wolfgang, M., et al. (1972). *Delinquency in a Birth Cohort*. Chicago: University of Chicago Press.

Yeudall, L., et al. (1982). "Neuropsychological Impairment of Persistent Delinquency." *The Journal of Nervous and Mental Disease* 170: 257–265.

Yeudall, L., et al. (1987). "A Neuropsychological Theory of Persistent Criminality: Implications for Assessment and Treatment." In *Advances in Forensic Psychology and Psychiatry*, ed. R. Rieber, vol. 2. Norwood: Ablex.

Ziskin, J. (1975). *Coping with Psychiatric and Psychological Testimony*. Beverly Hills: Law and Psychology Press.

Ziskin, J., and D. Faust. (1988). *Coping with Psychiatric and Psychological Testimony*. Venice: Law and Psychology Press.

12

Law, Culpability, and the Neural Sciences

Michael H. Shapiro

Abstract: The identification of neurochemical factors in social and criminal behavior does not necessarily undermine the traditional concept of legal responsibility. The philosophical and legal bases of existing concepts of culpability assume causality, and recent discoveries in neuroscience do not justify abandoning our presumption that the individual is responsible for criminal acts. Nevertheless, attitudes and beliefs may shift under the impact of the new information.

I. HOLD THE REVOLUTION: ARE THE NEURAL SCIENCES NEUTRAL WITH RESPECT TO FREEDOM AND RESPONSIBILITY?

A. In General

It is not news that the brain has something important to do with the way we think and act, but we have only within recent decades begun to identify certain neurochemical and neurophysiological correlates of mentation and behavior.[1] The resulting body of knowledge, though sharply limited, has worked major changes in our understanding of both normal and abnormal mental functioning and behavior and is accompanied (and in many cases stimulated) by developments in the pharmacological treatment of mental disorder. Such knowledge, however, is far short of providing us with a "book of life" on everyone's future.

Does this knowledge of the physical "substrates" of thought and action "force"[2] major changes in any conceptual tools we use to describe or evaluate actions and states of affairs? As Strawson asks, "What effect would, or should, the acceptance of the truth of a general thesis of determinism have upon these reactive attitudes [our normal, interpersonal "participant" attitudes, such as gratitude and resentment]?"[3]

179

Are we compelled to abandon ideas of freedom and responsibility[4] because the very functions these notions serve—for example, blaming and praising—no longer make sense, given our discovery of the biological antecedents of decision making?

I suggest a distinction—not a very sharp one—between conceptual changes required to preserve an idea's usefulness, at least under new circumstances, and changes in attitudes or beliefs stimulated by the sheer impact of new knowledge or technology. (In some cases, we may think that the concept must be discarded altogether because its presuppositions have been eroded.) Concerns about the meaning of "death" serve as an example of both. The ability to maintain cardiovascular function artificially while brain function (including that of the brain stem) is destroyed requires a more precise specification of the nature of death, but not a substantial expansion or contraction of the situations to which the term "death" is already thought to apply. The concept of "brain death" gives us guidance under new circumstances and better reasons for thinking and acting as we already do. (What is important is not so much breathing and heartbeat as the cessation of all capacity for mental and physical activity through loss of brain function.) The clarified idea of death thus helps define the legal powers and responsibilities of families, physicians, and the state and so allows us to work with the idea under changed conditions.

In contrast, the ability to keep severely brain-damaged persons organically alive (those in a persistent vegetative state, for example) does not require any conceptual change but moves us to rethink important values concerning the care of persons with severe cognitive impairments. Many observers have focused on the similarities between death and impairment that destroys nearly everything associated with one's unique identity (perhaps one's very personhood, on some views), and some have concluded that we should expand what "death" designates. Such a revision is not necessary for "death" to retain its historic usefulness. Nevertheless, the prospect of having to maintain the bodily functions of a human organism whose personality and consciousness are irretrievably gone is, for many, a powerful incentive to enlarge the meaning of "death" so as to allow termination of care and closure of a relationship.[5] Indeed, after reflection, some of us may come to view such an expansion as "necessary" for dealing with new circumstances.

As I said, the distinction between "forced" conceptual change and the rethinking of basic ideas and value conflicts is a loose one, and both processes may occur simultaneously—a necessary conceptual reconfiguration taking place through unavoidable normative debate. A forced change may thus require value rethinking. Consider the split between gestation and genetics through surrogate gestational motherhood (where a woman bears a child not genetically "hers"): we are driven to do major work on the idea of motherhood if it is to be of full service under new circumstances. (Which "mother" is entitled to custody? Even if we turn to "the best interests of the child" as a standard, the question of motherhood still commands our attention.) We are not simply specifying more precisely the elements of a concept we already think we know how to use, as with a brain function criterion of death, which represents a forced but limited change. What we are doing is rethinking the very foundations of the nuclear family, a process driven by gestational surrogacy's fragmentation of a "unified" biological process and its concomitant challenge to the social arrangements built around them. (See section IV, below.)

How does this distinction between "forced" and "stimulated" conceptual reworking apply to new findings in the brain sciences? It is not clear whether such findings force any conceptual changes in our ideas of freedom and responsibility. For several commentators, some aspects of these ideas may remain intact; other aspects may not. In any event, our accumulating knowledge is likely to stimulate rethinking, and this could result in shifts in attitudes and beliefs about the nature of choice and culpability.

The idea that determinism does or does not require conceptual revisions is of course controversial. It may be useful to distinguish between the effect of incremental knowledge when one already believes in determinism and threshold knowledge that persuades one in the first place of the grand view that human behavior is indeed caused and that determinism is true. If one already holds to determinism, the incremental knowledge should force no conceptual shifts (though it may stimulate conceptual rethinking). But if we are talking of threshold knowledge, the matter is less certain. Although some features of culpability assessment may remain intact even under deterministic assumptions, others may become frayed.[6] Dennett raises the issue of maximum destructive change in describing the free will problem as a "bogeyman"— a "dread secret" that "will paralyze us by shattering some illusion that is absolutely necessary for the maintenance of our lives as agents. Our own rationality will undo us, because once we've seen the truth, we will be unable to deceive ourselves any longer."[7]

I suggest, nevertheless, that our accumulating knowledge of brain chemistry need not have a broadly destructive effect (and no effect at all on some views) on the usefulness of moral or legal notions of responsibility. This is not a domain where ideas, to retain their full usefulness, must be transformed or discarded altogether because they rest on false assumptions. It is more probable that ideas like the "serotonin hypothesis"[8] can be viewed as neutral within the framework of freedom-responsibility debates—at least for important aspects of freedom and responsibility. If neurotransmitter chemistry or other aspects of brain functioning turn out to be correlated with what we *already* view as serious impairments of reason or affect, this does not directly add to the case for exculpation or mitigation; the *impairment* is what counts under an insanity or diminished capacity defense, not the underlying causes, physical or otherwise.[9] Even if someone's physiological system looks odd or abnormal in a statistical sense, this in itself is not a ground for exculpation: what must be determined is whether the person's forms of thought and feeling reflect relevant forms of deficit—in reasoning, self-control, and the like. Thus, if a neurotransmitter anomaly (or any other physical condition or process) is not connected with such impairment, it should in theory have no effect on culpability analysis (although one never knows what will happen in court when claims about physical anomalies are raised).[10] Similarly, finding no physiological anomaly does not defeat a claim that one is impaired.

Still, ideas of causation, freedom, and responsibility carry a strong emotional impact and may have important effects on how we characterize and deal with certain behaviors.[11] "Unforced" conceptual retooling may occur, through attitude shifts arising out of complex psychological processes.[12] The adage "to understand is to forgive" has pull, whatever its theoretical limitations. Though under prevailing culpability theory the exact nature of the cause of impairment is immaterial, there seems to be some connection we make between the very identification of physical correlates of undesired conduct and viewing such conduct as reflecting "impairment." With highly established correlations, then, we may take neurochemical findings that seem "anomalous" to be strong signs of severe impairment. We may characterize the thought and behavior in question as embodying disorder, partly because we have traced differences in how different persons are "wired up," and we compare these anomalies with what we think are confirmed examples of physically caused disturbances (e.g., drug intoxication, brain tumors). We may also rate a condition as more serious when the physical correlates suggest that it is nontransient, perhaps arising from a genetic anomaly or some enduring injury.

Ascriptions of impairment thus drive many arguments for exculpation and mitigation: known physical conditions that appear to be "different" seem to add to the "I couldn't help it" argument and inspire rethinking of our views about choice and responsibility. After identifying a distinctive physical condition, we may thus resolve doubts in favor of ascribing disorder and recognizing a claim of excuse. Perhaps the vividness of the physical condition partly accounts for this.[13] (Ascriptions of disorder and impairment arising from what are seen as physical causes may also lead more easily to administration of organic therapies.) All this holds even though the ascription of "disorder" and consequent impairment is not logically connected with its physical antecedents.

B. The Compatibilism Debate

I can summarize many of these points by returning to the earlier suggestion that both threshold and incremental knowledge of causation may be, at least in considerable part, conceptually and morally neutral:

•Those who believe causal determinism[14] holds and is inconsistent with moral responsibility obviously will continue to believe this when advised of the "details" of how causal determinism works.

•Those who believe causal determinism holds but is compatible with—perhaps required by—freedom and responsibility will be unimpressed with the idea that particular causal formats force them to abandon their views; they too have been assuming causal determinism all along.[15] They will say that the recent findings are merely particular instantiations of their general assumption[16] and are irrelevant to the freewill debate.

•Those who believe indeterminism holds and is consistent with or required by ideas of freedom and responsibility are not likely to be dissuaded by more scientific knowledge. They are presumably familiar, at a general level, with the workings of science. They know that there are physical correlates and causes of various aspects of human thought and action. This is perfectly consistent with indeterminism: no sane indeterminist believes that all the workings of the universe—including the actions of persons within it—are acausal. It may well be, however, that developments in the neural sciences will convert some marginal indeterminists to determinists through some complex process of attitude shifting. If they are indeterminists who think freedom requires an indeterministic world, they should, as new determinists, reject the possibility of freedom and responsibility.

In this way—a hard-to-specify psychological process—the new knowledge might cause a relative loss of support for the idea of culpability. One might view this as a forced conceptual change for a given person who is converted to determinism, but the variable is belief in determinism and in its supposed conceptual consequences, not in the idea of freedom, which the convert (mistakenly or not) already believes incompatible with universal causation. Dennett has obliquely suggested much the same: "Modern science isn't *making* determinism true, even if it is discovering this fact, so things aren't going to get worse, unless it is believing in determinism rather than determinism itself that creates the catastrophe."[17]

•As for those who believe indeterminism holds and is inconsistent with freedom and responsibility, they either will remain indeterminists or become determinists of the compatibilist persuasion—that is, determinists who believe it sensible to make ascriptions of freedom and responsibility.[18]

There are lots of "ifs" in the preceding account, but I am not going to embark on a study of compatibilism or free will generally.[19] I will just make a quick sketch.

Morawetz suggests that freedom requires a network of causality: "Whether one is the sort of person who is able to make decisions and exercise 'free will' is not a matter of whether one is able to alter one's personality and commitments wholesale [or to suspend causal constraints generally, one might add]; it is a matter of how one operates within them."[20] The underlying notion here is that freedom and responsibility are not only not inconsistent with causal determinism but also require it. As Nozick puts it, "[D]eterminism, rather than undercutting the subjunctives [propositions about what we could have done under various circumstances], may be their substratum."[21]

For compatibilists, then, as Flew notes, "the expression 'of my own freewill' refers to the absence not of causes but of compulsion."[22] In this context, there are no "uncaused causers," as many have noted.[23]

Compatibilism is nevertheless a vulnerable position. It may well be that confining the idea of "freedom" to matters of human control and compulsion (and perhaps to environmental constraints) is an unjustifiable conceptual move and that once our attention is called to matters of causation in general, the idea of freedom is ultimately incoherent: neither the compatibilist nor the opposing interpretations work.[24] I cannot pursue this any further. The point here is that specific incremental knowledge in the neural sciences does not directly affect the debate, though it may affect the numbers and identity of the partisans by producing converts to determinism.

C. The Criminal Law and the Idea of Mental Disorder

1. Mental Disorders and Opportunities for Control Within a Causal Framework Defined by Character

I suggested that much biophysical knowledge may be neutral concerning matters of moral responsibility. It should also be neutral concerning various forms of legal responsibility, fault-based or not. The constituents of responsibility in the criminal law are found in the defining elements of the offenses (e.g., knowledge, intention, purpose) and in certain defenses—those of excuse (e.g., insanity, diminished capacity, provocation, duress) and of justification (as in necessity or, on some views, self-defense). The idea of intentionally producing a result, for example, captures an intuitively plausible aspect of guilt. Duress and impaired mental capacities suggest plausible negations or mitigations of culpability, whether viewed as compromising our opportunities to choose freely or as keeping us from acting in accord with our characters.[25]

But suppose the terms of the discussion move beyond descriptions of mental states and into their physical substrates. If we say that one "cannot help" having certain intentions or character traits—they are fixed by features of our neurochemical systems—do we rob intention and character of their relevance for culpability and so erode culpability itself?

There are different forms of "cannot help" here, as the earlier remarks on the freewill debate suggest. One form is simply about causation: one's intentions are caused; one's very review of alternatives and evaluations of reasons are caused; and we cannot help but do what we are caused to do. Whether or not our possible reasons for action are sound, our very rationality in appraising them and using them in making decisions is itself locked into a causal network. This view is simply an element of the claim that causal determinism is incompatible with freedom and responsibility and does not rest on any particular forms of causation.

The other "cannot help" has to do with the nature of the causes. One has opportunities within a causal framework to perceive and assess reasons in deciding upon a course of conduct. These are not contra-causal opportunities—they are defined within a framework of causes.[26] But these opportunities are constricted when normal brain functions are impaired by some structural or electrochemical anomaly. A brain tumor that provokes certain forms of mentation establishes a subcategory of causation that is different from the norm. As Moore argues, "Causation is not compulsion. If we want to show that some causally relevant factor constitutes a compulsion, we can do so only by showing that that factor interferes with practical reasoning."[27]

We also judge our opportunities to be diminished when, without reference to physical (or other) causes, a person seems bereft of standard reasoning faculties or faces unusual difficulties in using them because of matters of mood or felt compulsions. "He couldn't help it" in this context is not a complaint about causation generally—indeed, it presupposes it. It is a complaint about particular sorts of causes that leave one unable to rely on rational faculties of thought, feeling, and control within a "standard" causal system.[28] Neurochemistry in general does not defeat normal reason and feeling: on the contrary, from what we know, it makes them possible. In fact, many argue that it defeats an ascription of responsibility to urge that someone has "escaped" the causal network defined by his or her character,[29] which is in turn mediated by brain functioning. To claim such escape is in effect to say that the preexisting capacities for thought and control within the causal framework embodied by the person's character have been impaired or lost.

Note that the discovery that some persons have nonstandard physico-chemical brain characteristics does not, standing alone, establish nonculpability. And nonculpability does not require identifying such characteristics. It is not necessary to connect physical anomalies to loss of opportunities for thought and self-control; a finding of nonresponsibility may be justified precisely because of the loss of controls, without reference to the supposed causes.[30] Physical anomalies are neither necessary nor sufficient for rejecting culpability.

2. Beyond Disease: The "Therapeutic State"

To view all crime as the product of disease is simply not required by any aspect of brain science.[31] One might abandon the traditional responsibility model on various moral grounds—for example, fairness or utility—but this is another matter. In any event, the general idea of a therapeutic state is nothing new,[32] and the problems it poses have been well discussed: the risk of injustice through indeterminate sentencing; describing as "disordered" any disapproved conduct (similar in outcome to abandoning notions of disorder altogether in favor of the behaviorists' "maladaptive behavior"); procedural laxity; compelled treatment; the erosion of the value of autonomy that may accompany weakening of responsibility notions; a loss of the deterrent and teaching power of the criminal law and criminal punishment; and a general inability to predict and guide one's conduct because of the difficulties in dealing with expansive notions of mental disorder or maladaptive behavior.[33] (This list is not meant to be exhaustive.) Some may argue that this is no worse than what the criminal justice system gives us in practice, but that cannot be addressed here. (On the therapeutic state's use of mass screening for certain behavioral propensities and on coercive treatment for mental disorders—or for "maladaptive behavior" independent of a disease model—see section II, below.)

D. Civil Law

Civil law is not a fertile area for assessing the impact of advances in biological psychiatry, though it is not wholly irrelevant. Fault-based tort law, with its emphasis on intention, recklessness, and negligence, does seem to talk the same culpability language as criminal law and expressly deals with "wrongdoing." Tort defenses suggest the same—for example, contributory negligence and assumption of risk. Finally, the foundations of tort liability include ideas about corrective justice and rectification through allocation of losses to culpable parties. Nevertheless, tort liability is also rationalized as an efficiency-promoting tool for encouraging due care and imposing costs on those best suited to reduce risks of harm. So, either the analysis collapses into what was said earlier about criminal liability, or it seems beside the point, given the functions of civil fault analysis.

Parallel remarks apply to contract law, which suggests fault analysis in allowing for punitive damages and for various excuses for nonperformance—for example, undue influence or coercion. One might argue here, as in criminal law, that undue influence is indistinguishable from the causal constraints of electrochemistry. But contracts analysis is not a distinctively promising avenue for illustrating the effects of greater knowledge about brain chemistry or molecular biology.

II. PREDICTION, PREVENTION, AND TREATMENT: NORMATIVE REEVALUATION OCCASIONED BY NEW TECHNOLOGICAL POSSIBILITIES

A. *Discovery and Detention: New Forms of "Guilt" or the End of Guilt?*

Knowledge that explains is kin to knowledge that predicts and knowledge that facilitates control. Ideas like the serotonin hypothesis suggest the possibility of being able to predict with some precision who will engage in what conduct and when. What are the conceptual and normative consequences of those technological possibilities? If we are not forced to abandon or significantly revise our ideas of responsibility, will we be moved to rethink them in light of the new possibilities for *effective* prediction and control? If we were to decide, for example, that a propensity for violence or other unwanted conduct was itself "culpable" (an aspect of character?) and worthy of criminalizing, that would be a major alteration of our notions of responsibility, which focus more (but not exclusively) on actions than on statuses or proclivities.[34]

Although these results are clearly not entailed by new knowledge (it is not forced in the sense I suggested earlier), we have never fully confronted the issues because they have not come up as realistic possibilities.[35] But if we actually knew that certain persons were very likely to act in unacceptable ways, the impact of this knowledge and its practical crime prevention opportunities might tilt us toward giving up the idea of assigning guilt on the basis of actions rather than evil tendencies. (We are, after all, familiar with the idea of original sin.) Or, as I said, we might simply view guilt as irrelevant to matters of public safety. Either way, the result might be preventive detention.

The question of the irrelevance of guilt bears further attention. Whether or not we accept a notion of behavioral propensity as guilt, we are faced with the practical question of what to do with those identified as likely lawbreakers and with the threshold question of whether to identify them in the first place. Those who would opt for compulsory mass screening to determine, for example, a profile of the electrochemistry of each person's brain or the base sequences in each human genome would likely assert that matters of guilt and responsibility are beside the point. Protecting the public against known, significant probabilities of harm by detaining and treating prospective harmdoers has no connection with ideas of responsibility, or so the argument might go. It is simply a matter of social self-protection, to be weighed against other values. Indeed, one might envision the recognition of a community duty to set up screening, detention, and treatment programs and to recognize individual duties to submit to them.[36]

Compulsory screening and preventive detention have of course not yet been generally embraced as appropriate practices.[37] Perhaps this is partly because the scientific foundation for them is not yet in place. But current indifference or hostility to screening and detention occurs in a context of limited predictability. What one thinks of a program believed impossible to implement is one thing; what one thinks when the program seems realistic may be quite another. The social-psychological impact of having strong predictive and therapeutic abilities may shift our ideas about the duties that individuals owe to the community (to submit to screening and other measures) and that the community owes to individuals (to prevent predictable harms). If we are unable to predict or control behavior, it is easy to make lofty claims that justice requires action as a precondition for responsibility and that status or predisposition are insufficient. But technology may call our bluff.[38] While no theoretical necessity compels it, technological capabilities invite development of justifications for their use. We are easily led to " 'can' implies 'ought.' "

This, again, is not a forced conceptual change—one that requires restructuring a concept to maintain its usefulness or even discarding it because its presuppositions are eroded and its usefulness thus impaired; it is a change in attitude arising from a change in circumstances that provokes value analysis. We *may* come to feel differently about guilt or preventive detention in a world of new knowledge and capabilities— some may even forecast the death of criminal law as we know it—but we do not *have* to. Determining the mechanisms of such attitudinal change is the domain of cognitive psychologists and allied investigators.[39]

B. Coerced Organic Therapy

Certain psychotropic drugs seem to be highly effective in treating mental disorders.[40] There are strong pressures to permit the coercive use of these therapies on confined mental patients and on prisoners (even nondisordered prisoners). Indeed, the U.S. Supreme Court has already ruled, in *Washington v. Harper*,[41] that, where a convict's dangerousness or disability is attributable to mental disorder, involuntary administration of antipsychotic drugs may be permissible without a judicial hearing, where adequate alternative procedures are in place. The state's interests were held to outweigh the "liberty interest" in avoiding involuntary treatment with antipsychotic drugs.[42] Such therapeutic agents are not "magic bullets": they do not simply "delete" particular propensities to think and act in specified unlawful ways.[43] If they "neutralize," they often do so through their nontherapeutic side effects.[44] But technological developments could theoretically provide more precise mechanisms for controlling behavior.[45] If this occurs, the practical realities of detailed behavior control would push us to confront the security versus liberty conflict (and other conflicts as well). It will not do simply to say that we could choose to avoid such measures: the very existence of an option puts a strain on decisionmaking and thus creates "transactions costs." Perceived choice, then, can be burdensome just by being there. As with prediction and preventive detention, the live prospect of effective "treatment" (whether in conjunction with a disorder model or not) prompts us to reevaluate the balances we have so far struck in our value conflicts.

C. Augmentation

Later, the conflict may become still more complex: knowledge of the brain's neurochemistry may make it possible to alter and enhance traits selectively—various sorts of intellectual aptitudes, for example.[46] The model for action here is augmentative, not curative.[47] In the face of such striking technological powers, there may be a revision of the idea of identity because its presuppositions are eroded. "Person" may, for better or worse, come to refer to a set of potentials for manipulation rather than a relatively fixed set of characteristics whose integrity we are to respect.[48] Even with less impressive powers, however, the very prospect of altering traits suggests the malleability of human identity and might lead to reordering of value hierarchies and recasting of conceptual tools. Indeed, the decades-old debate about the propriety of the therapeutic state reflects such nascent technologies.

III. CONTROL OVER KNOWLEDGE OF A PERSON'S PHYSIOLOGICAL STATE OR GENOME

Consider the purposes for asserting control of information about someone's physical or genetic makeup. Social defense against unwanted behavior and encouragement of desired behavior are obvious ones. Employers, both government and private, may wish to promote work efficiency and reduce liability risks by knowing the susceptibilities to disease or injury and the behavioral predispositions of present or prospective employees. Responsive action may include preventive medical care, transfer, termination, and refusal to hire. Insurers also have obvious interests in such information. The project to map and sequence the human genome may thus generate calls for determining the individual genomes of all persons who are or may be employees or insureds—practically everyone.[49]

Despite recent U.S. Supreme Court decisions eschewing talk of privacy and fundamental rights, it still seems appropriate, when dealing with physiological or genetic profiles, to resort to the idea of privacy as at least an important "liberty interest" (the Court's currently favored term of art). When we focus on the idea of the body as a repository of information that can be communicated to others for various purposes (commercial gain, criminal prosecution), privacy and confidentiality notions seem intuitively plausible, as when they are invoked against law enforcement authorities or insurance companies.[50] (This does not, however, exclude the parallel use of property ideas.[51]) The constitutional question, in general form, would be whether the state has an interest strong enough to override one's preferences for privacy.[52]

A recent example of commercial and scientific use of biological information derived from a particular individual is the "Mo cell line" case, where researchers allegedly took tissue from a patient suffering from hairy cell leukemia without disclosing that its proposed uses might be financially rewarding. What was valuable was not so much the physical stuff of Mr. Moore's spleen cells but certain genetic information they contained. The California Supreme Court held that "the allegations state a cause of action for breach of fiduciary duty or lack of informed consent" but rejected a theory that the patient's property was converted.[53] Here, notions of "confidentiality" do not seem apt, although "privacy" in the sense of controlling the fate of one's physical person and its constituent parts might apply. The case is also a reminder that protection of various forms of personal integrity can come from sources of law other than the U.S. Constitution—a state's common law, for example.

IV. AN ANTHROPOLOGICAL VIEW: FURTHER COMMENTS ON THE CONCEPTUAL AND NORMATIVE IMPACTS OF EXPLAINABILITY, PREDICTABILITY, AND CONTROLLABILITY OF MIND AND ACTION

A. Effects of Human Self-Knowledge

I raised earlier the distinction between forced conceptual changes and the pressure to rethink value issues in the light of new knowledge or technique. Some of the examples bear further explanation.

1. Conceptual Reconstructions Forced or Inspired by the Ability to "Fragment and Reassemble" the World

The distinction I made was between having to reconstruct concepts to retain their usefulness under changed conditions or even to abandon them after their presuppositions are eroded, and reevaluating concepts and value collisions upon being confronted by new options and new knowledge that put certain issues "in italics." There is no clear border between the two, and both processes may be going on simultaneously.[54] The idea that concepts of freedom and responsibility must be scrapped because of changes in our understanding of their presuppositions was raised and criticized earlier when discussing freedom and responsibility. Either the asserted presuppositions are not presuppositions, or, if they are, they are not refuted by new knowledge in the neurosciences. But other concepts might indeed require "rational reconstruction"[55] or abandonment (at least in certain contexts). For example, the idea of personal identity loses some descriptive and guiding value when human traits are significantly malleable. A presupposition of the idea of identity is that traits do not just melt away or transmute before our eyes.[56]

More generally, the ability to fragment and reassemble various life processes may challenge our categories for describing and evaluating conduct and states of affairs—the very tools we use in much of our everyday thinking.[57] There is no shortage of examples of "separating" the constituents of life processes. As we saw, the concept and criteria of "death" require some attention when neurological and cardiovascular failure are separated by life-support technologies; the usual outcome is a "brain death" statute.[58] The idea of death is also challenged by the growing capacity to separate organic life from life as a functioning person: we can maintain organic life long after life as a conscious person has irreversibly ended. This power pulls some observers toward revising "death" so that it focuses on "personhood life" rather than "organic life."[59] The latter seems to represent an expansion in the idea of death rather than a deeper account or further specification of its existing meaning.[60] So, too, "mother" becomes problematic with the separation of gestation from genetics: one woman can gestate the genetic offspring of another woman—a fragmentation that is accompanied by an obvious social fragmentation in which reproduction crosses marital or other familiar associational lines. Finally, "merit" requires rethinking when valued attributes are pharmacologically variable (say, by intellect- or strength-enhancing drugs), enabling us to focus on a particular trait, partly "isolating" it from its developmental context, then altering and reintegrating it into the person.[61]

2. Value Reordering and Conceptual Shifts When New Options Arise

When certain options are difficult or impossible, we can comfortably ignore them and the moral and legal issues they raise. But when confronted with certain technological possibilities, we might, as suggested earlier, feel pushed into reinvestigating and reinterpreting a variety of ideas; our attitudes may shift, and we may resolve some value collisions differently. As I argued, such revision is not necessarily required by changed conditions, but it may be provoked by confrontation with certain matters that suggest the redrawing of normative maps. This was suggested earlier in the discussion of screening and "treatment" for violent propensities: the way we balance liberty and security may be altered, both practically and theoretically, when we *can* predict and control behavior. Other contexts suggest other examples. If we can save lives through fetal tissue and organ transplants, for example, will we rethink the idea of fetal value, downgrading it further, and eventually endorse the conception and abortion of fetuses for medical use?[62]

As Wiggins suggests, "[A] man's reflection on a new situation that confronts him may disrupt such order and fixity as had previously existed, and bring a change in his evolving conception of the point . . . , or the several or many points, of living or acting."[63]

3. Demoralization Effects[64]

a. *Nonuniqueness: Continuity with other entities and processes.* As was suggested long ago by Mazlish, scientific knowledge may erode the sense of "discontinuity" we perceive between ourselves and machines, animals, and the external world generally.[65] The more we investigate, the more we see the smooth lines from ourselves to other forms of life (animals have neurophysiological systems too), from ourselves to machines (if we are discrete causal systems, as are machines, we are linked to them in a fundamental way), and from ourselves to the universe itself, viewed as a machine, an organism, or an information processing device. To be able to screen, predict and control humans seems to strengthen these perceived connections. Seeing continuities and links where before we saw vast gulfs does not always necessitate conceptual changes that affect values, but it may stimulate new conceptual and normative structures that adjust our views of our place and entitlements in the world. In general, the vivid ability to separate and rejoin various aspects of biological processes through recently developed technologies may emphasize the haziness of the borders between ourselves and everything else. If such borders vanish, do we assimilate other things to ourselves, or do we see ourselves as like those other things—or both? It is this persistent question that drives fears of "commodification" and "objectification"—of our descent from persons to objects.[66]

b. *We see things when they fail: The revelation of deficits in belief and value systems.* Things that operate smoothly, even if not ideally, are often not noticed. And if we do not notice them, we may not notice all their deficiencies, either. But when there is a serious breakdown in a system whose smooth operation has blended into a barely noticed background, we see the system and its failures.[67] For example, our system for distributing important commodities is challenged during acute shortages, say, of lifesaving resources.[68]

Similarly, when we are faced with the prospect of screening for predispositions toward unwanted behavior, or the possibility of human reconstruction and behavior control, or the ongoing "artificial" separation of historically unified aspects of reproduction, we see the limitations of our current tools of analysis. Accurate biophysical screening to predict future conduct has only limited parallels in prior practice.[69] Our ideas about autonomy, privacy, justice, and fairness do not yield ready answers about whether such screening, voluntary or involuntary, is permissible.

The possibility of human reconstruction (or "demolition," to borrow Alfred Bester's term)[70] also goes far beyond past efforts at behavior control. Existing normative tools suggest that involuntary demolition is impermissible. But the matter is less clear when it is voluntary.[71] Our very awareness of the option may make us review whatever principles we think limit or authorize such uses of technology. New choices, even if they compel no conceptual reconstruction, may thus lead us to alter our preexisting "resolutions" of value conflicts.[72]

A byproduct of such reconstructive processes may be demoralization: the revelation of limits and infirmities in our normative systems can cast doubt on their overall soundness. (When the clock strikes thirteen . . .) This may not be a sufficient reason to avoid technological change—indeed, new concepts and evaluations may be all to the good and well worth the discomfort.[73] The breakdown of certain principles and rules will not necessarily result in thoroughgoing skepticism about an entire belief system.[74] But if demoralization and loss of confidence in our systems of thought are "costs" of advancing knowledge and technology (it may be controversial whether these effects occur or are indeed costs), we ought to be aware of them.

4. Objectification and Commodification: More on Fragmenting and Reassembling the World

We saw that certain technologies enable us to "divide and conquer" life processes and to reassemble them in new ways. Such measuring, tinkering, and rearranging is characteristic of what we do with *things* (living and nonliving) and with technological or commercial systems. Some observers thus see a risk that the ever-more-clearly-perceived "continuities" between persons and things will be emphasized, thus providing us both with a slope and a downward acceleration into a lower world inhabited by human objects rather than by human persons. Prediction and control of mind and behavior are obvious examples of these possibilities, although they also suggest possibilities for enhancing individuality and human capacities, perhaps distancing us from the world of things and objects.[75] The point, in short, is that the possibility of reassembly at our direction creates risks of objectifying ourselves and commercializing our relationships and interactions. Such risks are, I think, often exaggerated, but they need to be taken seriously.[76]

A connected risk is that an ethic of perfection will come to dominate certain technologies. If so, certain attitudes toward others that are now (at least in our mythology) "noncontingent"—love for our children, respect for the separate identities of others—may become contingent on our success or failure in manufacturing the traits and persons we wish and on the ultimate success of the persons we produce.[77]

B. Too Much Choice?

Consider again the choices opening up: to screen populations in order to predict behavior or susceptibility to illness, to control mind and behavior, to alter the human genome. In raising the idea of too much choice[78] I am not attacking technological development; I am simply noting the possible costs of enhanced choice. Although we ordinarily prefer more choice over less, some options may make us normatively worse off—in certain respects, not necessarily overall—for several reasons. (a) They create a risk of culpability for failing to act in ways that were technically or legally impossible before. Think of someone in a low-income family who, when advised that selling organs has been legalized, concludes that he or she may be morally obliged to sell one to help the family financially.[79] (Of course, he or she has a chance to be heroic, and may prefer the expanded set of opportunities.) More to the point for our purposes, think of failure to discover violent propensities and to act on them: might this be a culpable omission given new technological powers to forecast and control? Is there a community duty to unidentified potential victims of violence to reduce the risks by screening and treating the violence-prone? (b) The new options may, as suggested before, also cast doubt on the worth and utility of an overall value system by displaying its deficits—its inability to cope with certain new options.

To increase opportunities is, other things being equal, to enhance autonomy—or so one might think. We can always decline the choice, so we are clearly better off, aren't we? Perhaps so, but we should check out the costs before concluding that new choices make us better off, overall. To "decline" a choice requires effort and may be the product of painful reviews of the most basic ways we have of describing and judging ourselves. More choice may conceivably impair autonomy and adversely affect other values. That is why some will be happy if we never learn anything more about neurotransmitters and their effects on behavior.[80]

ACKNOWLEDGMENTS

My thanks to Elyn Saks, Scott Altman, and Roy G. Spece, Jr., for their helpful comments.

NOTES

1. See generally S. Snyder, Drugs and the Brain (1986). Bear in mind, however, the risk of overstating the degree of knowledge and control we have. In particular, "correlations" carry us only so far and do not necessarily establish causal lines. Similarly with "substrates," a term I will use loosely. See generally H. van Praag, 176 J. Nervous & Mental Diseases 195, 196, 198 (1988):

> [B]iological psychiatry has provided psychiatry not only with a new basic science and new treatment modalities, but also with the tools, the methodology, and the mentality to operate within the confines of an empirical science. . . . Psychiatry is on its way to becoming a scientific discipline. . . . On the other hand, it must be acknowledged that some of the

original expectations have remained unfulfilled and that the biological revolution has also generated undesirable side effects.

We have discovered few new drugs recently. . . . Only a few new therapeutic principles have been added. . . .

So far this search [for biological markers] has resulted in few, if any, findings with practical/diagnostic value. The hope of finding objective signs (markers) of psychiatric diseases . . . has not yet been fulfilled. Nonsuppression of cortisol after dexamethasone, blunted thyroid-stimulating hormone response after thyrotropin-releasing hormone, low cerebrospinal fluid concentration of 5-hydroxyindolacetic acid, low platelet monoamine oxidase, to mention only the most obvious examples, have been introduced as diagnostic procedures but have all turned out to be of low specificity and to occur cross-diagnostically. . . . Because a particular psychological dysfunction is seldom specific for a given syndrome or nosological entity—occurring rather across diagnoses—this would explain the cross-diagnostic occurrence of most biological 'markers.' A case in point is the evidence of disordered serotonin function, which appears to be related not to any disease entity or syndrome, but to disordered aggression regulation irrespective of diagnosis. . . . The final danger [in diagnosis] is the tendency to inflate the explanatory and therapeutic weight of biological findings.

Moreover, we might not even know the direction of causality in given cases.

See also van Praag, *Biological Suicide Research: Outcome and Limitations*, 21 Biol. Psychiat. 1305, 1317 (1986) (his summary of the article): "Most available evidence suggest a relationship between diminished 5-HT [serotonin] metabolism and suicidal behavior. The hormonal data are so far inconclusive. Apart from summarizing the present state of the art, methodological flaws in biological suicide research are discussed. . . ." He also points out, in discussing methodological problems, that "[d]ifferent neural circuitries and neurochemical patterns are associated with different forms of aggression. . . . The nonunitary nature of human (auto)aggression is obvious. . . . [T]he driving forces behind suicidal acts are complex and so is the underlying emotional status. Hence, it is not warranted to link biological findings in suicide victims simply to disturbed aggression regulation. . . ." [*id.* at 1315]. And see M. Virkkunen et al., *Psychobiological Concomitants of History of Suicide Attempts among Violent Offenders and Impulsive Fire Setters*, 46 Arch. Gen. Psychiat. 604 (1989).

The assumptions and workings of biological psychiatry of course reflect a more general set of claims. See generally M. Shapiro & R. Spece, Bioethics and Law: Cases, Materials and Problems 11 (1981).

The "biophysical thesis" . . . holds that in theory we can explain in physical terms all life processes, including mental functioning, behavior, and physical attributes. A corollary of this proposition is that in theory we can by physical means control these life processes (whether or not we have full explanations for them or for how our interventions work). The thesis and its corollary hold both for "normal" as well as "abnormal" processes and traits.

See also the review of "biochemical and physiological findings as predictors of response to treatment" in W. Potter et al., *The Pharmacologic Treatment of Depression*, 325 New Eng. J. Med. 633, 635 (1991). And see J. Raloff, *Biochemical Aggression—The Legal Dimensions*, 124 Science News 173 (Sept. 10, 1983). For general discussion of biological causes of behavior, see Biology, Crime and Ethics: A Study of Biological Explanations for Criminal Behavior (F. Marsh & J. Katz eds., 1985); Biosocial Bases of Criminal Behavior (S. Mednick & K. Christiansen eds., 1977). See also P. Reilly, *Reviewing Proposals to Study Biological Correlates of Criminality*, 13 IRB 8 (Nov.-Dec. 1991).

2. In the text I distinguish "forced" conceptual revision from contingent normative reevalua-

tion. Forced revision may be induced by technological changes that make such revision necessary to retain the concept's usefulness under new circumstances, or even require its abandonment if its presuppositions no longer hold. (Forced conceptual changes may thus range from limited revision or specification to annihilation.) Unforced reevaluation may arise from the perception of new options for action, which may move us to reexamine prior value conflicts and alter our attitudes and beliefs. These two processes may overlap in given cases. Indeed, some forced changes necessarily involve rethinking of values and attitudes. I do not try to define "forced," "conceptual revision," or "conceptual change" more precisely. Note that "conceptual revision" does not necessarily entail "change of meaning." It may, for example, involve a new theory about the nature of what a term or concept refers to (e.g., the nature of death, the molecular structure of water). See *infra* note 5.

3. Strawson, quoted in D. Dennett, Elbow Room: The Varieties of Free Will Worth Wanting 15 (fn.) (1984).

4. I will not explore the connections between moral and criminal law ideas of freedom and responsibility.

5. It may sometimes be difficult to distinguish between a "further specification" of the criteria for the use of a term and an "expansion" or "contraction" of its coverage. See M. Moore, *A Natural Law Theory of Interpretation,* 58 S. Cal. L. Rev. 277, 294, 324, 326 (1985):

> We will not have changed the meaning of "death" when we substitute one theory for another, because by "death" we intended to refer to the naturally occurring kind of thing, whatever the true nature of the event turned out to be. . . . [O]nce [people] learned the facts about revivability, imagine their surprise at a legal decision that refused to even look at the facts about revivability as it authorized the cutting out of a "dead" person's heart. Most people would not only be surprised but shocked by any such conventionalist interpretation of "death." They would be shocked because they can fairly expect their courts to give "death" the meaning they themselves would give it: as the name of a natural kind of event about which we can learn all sorts of surprising facts without changing the meaning of the word at all. . . . The brain dead notion has supplanted the heart stoppage notion because the former is part of a better theory of death.

Much the same applies to the discovery of the hitherto hidden molecular structure of, say, water. H. Putnam, Reason, Truth and History 22–25 (1981). Compare Quine's discussion of Carnap's idea of "explication": "Any word worth explicating has some contexts which, as wholes, are clear and precise enough to be useful; and the purpose of explication is to preserve the usage of these favored contexts while sharpening the usage of other contexts." W. Quine, From a Logical Point of View 25 (1961). See also the discussion of "motherhood" in section IV; are we clarifying or significantly altering the meaning of "mother" if we hold that the term covers nongenetic gestational mothers, or genetic nongestational mothers, or both?

6. Honderich argues that "[w]e do not have a single conception of the initiation of action, or a single belief as to the role of such a conception. Our circumstance is not either that a determinism leaves moral approval and disapproval untouched, or that it destroys it. To suppose that it destroys it, as Incompatibilists do, is to ignore our attitudes which may issue in intransigence. To suppose that a determinism leaves moral approval and disapproval untouched, as Compatibilists do, is to ignore our attitudes which issue in dismay." T. Honderich, A Theory of Determinism: The Mind, Neuroscience, and Life-Hopes 539 (1988). He argues that both compatibilisim and incompatibilism are thus false. See his discussion at 486–487 for further elaboration of his view that some sets of attitudes and beliefs do require revision if determinism is true, while others do not. He criticizes Berlin's "wider view," expressed in I. Berlin, Four Essays on Liberty 113 (1969), Berlin argues that if "social and psychological

determinism" were accepted, "our world would be transformed more radically than was the teleological world of the classical and middle ages by the triumphs of mechanistic principles or those of natural selection. Our words—our modes of speech and thought—would be transformed in literally unimaginable ways; the notions of choice, of responsibility, of freedom, are so deeply embedded in our outlook that our new life, as creatures in a world genuinely lacking in these concepts, can . . . be conceived by us only with the greatest difficulty."

See also A. Flew, Crime or Disease? 95–115 (1973) in a chapter called "Determinist Presuppositions Cannot Disprove." He seems to separate three senses of "freewill"—one related to the causal principle, another to compulsion and control, and a third to being able to pursue one's desires. For the second and third senses, "it is irrefragably certain—certain beyond any possibility of upset by any other discoveries, whether in psychology, or in physiology or even in theology—that in both these senses freewill is a reality." (*Id.* at 100–101.) And as to the first sense (all events have "causally sufficient conditions"), "any such Freewill and a universal general Determinism must be incompatible. In that case there would be no philosophical problem; only the question whether Freewill, in that sense, in is fact an actual phenomenon. . . . [N]o discovery, simply as an unapplied discovery—not even the most supercosmic discovery—could ever abolish familiar differences [say, between acting under compulsion and acting of one's own freewill (*id.* at 96)]." (*Id.* at 100, 103.) Later, he states that "although familiar differences [as between external compulsion and self-direction] cannot be abolished by discoveries, such discoveries may sometimes demand some revision in accepted ideas about the significance of those differences." (*Id.* at 106.) The remark seems to use "demand" in the sense of "force" rather than "inspire" or "stimulate."

Finally, see V. Weil, *Neurophysiological Determinism and Human Action,* 89 Mind 90, 95 (1980) (discovery of neurophysiological mechanisms essential to our acts can augment our understanding of human actions without diminishing their significance in our moral life).

7. Dennett, *supra* note 3, at 7, 14. See section IV below for other examples of concept-threatening knowledge. He also notes the view that "[i]f determinism is true, it seems, there is no elbow room left for our selves, and no work for our selves to do" (*id.* at 13).

8. See generally Snyder, *supra* note 1, for discussions of the operation of neurotransmitters in the brain.

9. See section I-C, below.

10. Note the "XYY defense" cases. The effects of the supernumerary Y chromosome have not been fully traced, and the defense is unlikely to succeed. See, e.g., People v. Yukl, 83 Misc.2d 364, 371, 372 N.Y.S.2d 313 (1975) ("While there is strong evidence which indicates a relationship between genetic composition and deviant behavior, the exact biological mechanism has yet to be determined"); People v. Tanner, 13 Cal. App. 3d 596, 91 Cal. Rptr. 656 (1970) (similar). See generally W. LaFave & A. Scott, Criminal Law 377–382 (2d ed. 1986). The court in People v. Yukl refers to a French case in which evidence of an XYY condition apparently was useful in mitigating a sentence. 83 Misc.2d at 371.

Consider also what the connection may—and should—have been between the acquittal of John Hinckley as not guilty by reason of insanity and the CAT-scan showings of what were believed to be certain abnormalities in his brain. L. Caplan, The Insanity Defense and the Trial of John W. Hinckley, Jr. 79–85 (1984); S. Taylor, *Hinckley Is Cleared but Is Held Insane in Reagan Attack,* N.Y. Times, June 22, 1982, at A1, col. 6. Compare these remarks on an Australian XYY case in which the defendant was found not guilty by reason of insanity—but (as the authors argue) not necessarily because of the XYY anomaly: "It is . . . quite possible that the jury would have returned the same verdict on the basis of his mental deficiency alone, his abnormal electro-encephalograph and his epilepsy alone, or both, without there being any mention of his genetic make up." A. Bartholomew & G. Sutherland, *A Defense of Insanity and the Extra* Y *Chromosome:* R. v. Hannell, 2 Aust. & N.Z. J. Crimin. 29, 36, (1969), quoted in LaFave & Scott, *supra* at 379 n. 17.

11. See references in note 7, *supra*.

12. See generally H. Margolis, *Patterns, Thinking, and Cognition* (1987).

13. We may also raise questions about the identity of the person who seeks to be excused.

14. There are varying ideas about the nature of determinism. See generally the various works cited here (Dennett, R. Nozick, M. Moore, Wolf). Note the idea of "ignorance determinism," as Moore puts it, in which someone, though conceding that causes are operating and may one day be identified, "might urge that what determines an actor's responsibility is not the degree to which the actor is caused to act, but rather the degree to which *we* have knowledge of the causes of action. If studies showed us, for example, that eighty percent of the children raised in a certain environment commit crimes, we would excuse them from responsibility" (Moore, *Causation and the Excuses*, 73 Calif. L. Rev. 1091, 1118–1119 [1985] [emphasis in original]). But if we indeed believe in full causation, and further believe this to be incompatible with freedom, our ignorance of the causes is irrelevant, as Moore points out (*id.* at 1119). For his discussion of "partial determinism," see *id.* at 1114–1121. There is a problem of fitting quantum theory and its probabilistic notions into the idea of causation, but I will not discuss this except to say it is uncertain how any of this applies to behavior and macro-events generally. See *id.* at 1114–1118.

15. Some supposed compatibilists may be indeterminists without being fully aware of it; however, the pressure of new findings may force them to confront their views.

16. See *supra* note 1 on biological psychiatry and the biophysical thesis.

17. Dennett, *supra* note 3, at 15 (emphasis in original).

18. On compatibilism and incompatibilism, see Honderich, *supra* note 6, at 451–487.

19. See generally Dennett, *supra* note 3; R. Nozick, Philosophical Explanations 291–397 (1981).

20. T. Morawetz, The Philosophy of Law 190 (1980).

21. Nozick, *supra* note 19, at 326.

22. Flew, *supra* note 6, at 104.

23. Cf. Gerald Dworkin's critical reference to the idea of an "uninfluenced influencer." G. Dworkin, *The Nature and Value of Autonomy* (unpublished), quoted in R. Morison, *The Biological Limits on Autonomy*, 14 Hastings Center Rep. 43 (Oct. 1984). See also the critique of the idea of the "self-causing cause" and of "originators" as "causal" in Honderich, *supra* note 6, at 199–202.

24. Honderich, *supra* note 6, at 484–485, rejects the idea that our historic problems in integrating determinism into our conceptual systems are simply the product of long-standing confusion. He criticizes both compatibilism and incompatibilism.

25. On "choice" versus "character" perspectives on culpability and excuse, compare M. Moore, *Choice, Character, and Excuse*, 7 Soc. Philos. & Pol. 29, 58 (1990) ("We excuse because culpable choice is lacking, not because the action fails to manifest bad character"), with P. Arenella, *Character, Choice and Moral Agency: The Relevance of Character to Our Moral Culpability Judgments*, 7 Soc. Philos. & Pol. 59, 83 (1990) (if role of character is minimized in favor of "blaming the rational choices of disembodied rational agents, we will create a moral theory that neither describes nor justifies our actual blaming judgments").

26. See Flew, *supra* note 6, at 107:

Nor, since we are thus rational animals, can it be right to take physiological accounts of what goes on when I argue for some conclusion, or decide to perform some action, as precluding the possibilities, that I have good reasons for what I am saying, and that I shall do what I shall do because I want to. If this is so, and it surely is, [then] it must be wrong to say: "If human actions were completely explicable according to physical scheme of explanation, then they could not really be explained according to a rational or human scheme of explanation"; and wrong too to take Determinism to be self-discrediting on the grounds that the Determinist "does not hold his Determinist views because they are true, but because he has such-and-such a genetic make-up, and has received such-and-such stimuli."

27. Moore, *supra* note 14, at 1131. He continues: "Nothing in [the] status conception of insanity implicitly refers back to causation. It is not because their mental disease causes the insane to commit crimes that we excuse them, no more than it is because an infant's lack of rationality causes him to do bad that we excuse him. Rather, in both cases, we excuse because the actors lack the status of moral agents. . . . Our practices of assessing merit and responsibility are consistent with the proposition that persons are responsible for their (determined) choices, inconsistent with its negation." *Id.* at 1137, 1144. See also the section entitled "Does Neurophysiological Reduction Undercut Tracking [bestness or rightness]?" in Nozick, *supra* note 19, at 332.

28. For discussion of the insanity defense in its varying forms and of other nonresponsibility defenses, see LaFave & Scott, *supra* note 10, at 302–403.

29. "[I]f we require an agent to be psychologically undetermined, we cannot expect him to be a moral agent. . . . [V]iews that offer conditional analyses of the ability to do otherwise, views that, like mine, take freedom to consist in the ability *to be determined* in a particular way, are generally compatibilist views" (S. Wolf, *Asymmetrical Freedom*, 77 J. Philos. 151, 153, 162 [1980] [emphasis in original]). See also Flew, *supra* note 6, at 107, quoting Dostoyevsky's character Mrs. Marmeladov in *Crime and Punishment:* "She can't help herself, I'm afraid. [It's] her character, you see."

30. See generally the materials in S. Kadish & S. Schulhofer, Criminal Law and Its Processes 965–1016 (5th ed. 1989). There is, therefore, no "final" list of conditions that can make a case for formal exculpation or mitigation or for penalty reduction. It is possible that new understandings of certain forms of behavior will reveal them to reflect impaired cognition or control functions. See, e.g., J. Horney, *Menstrual Cycles and Criminal Responsibility,* in Marsh & Katz, *supra* note 1, at 159; Lee, *Postpartum Emotional Disorders,* Med. Trial Technique Q. 286 (1984). Neither the menstrual cycle nor postpartum defenses has enjoyed much success. Postpartum psychosis seems to be more readily acknowledged as a clinical entity by psychiatrists, but not as a significantly exculpatory one. See D. Trigoboff, *Postpartum Blues: Cases Test Use as Murder Defense,* L.A. Daily J., Dec. 16, 1987, at 1, col. 4.

Compare the remarks on excuse in J. Wilson & R. Hernnstein, Crime & Human Nature 504–505 (1985): "If society should not punish acts that science has shown to have been caused by antecedent conditions, then every advance in knowledge about why people behave as they do may shrink the scope of criminal law. If, for example, it is shown that sex offenders suffer from abnormal hormones combined with certain atypical relations with their parents, then, by the existing standards of responsibility, why should their attorneys not demand acquittal on grounds of bad hormones combined with a particular family history?" But, as suggested here, "antecedent causes" do not necessarily foreclose responsibility. Freedom and excuse, at least on some views, are interpretable within a causal framework. Wilson and Hernnstein in fact defend ideas of punishment and responsibility by focusing on excusing conditions. "Scientific explanations of criminal behavior do, in fact, undermine a view of criminal responsibility based on freedom of action. And it is also correct that this book has taken pains to show that much, if not all, criminal behavior can be traced to antecedent conditions. Yet we view legal punishment as essential, a virtual corollary of the theory of criminal behavior upon which the book is built. [They then refer to H.L.A. Hart in denying that this creates a paradox.] An act deserves punishment, according to the principle of equity, if it was committed without certain explicit *excusing conditions*" (emphasis in original) (*id.* at 505). Despite their comments on freedom of action, their position seems to be a compatibilist one.

31. It may also draw attention from social and economic conditions that may be predisposing conditions for crime.

32. For brief accounts, see Shapiro & Spece, *supra* note 1, at 171–174; Kadish & Schulhofer, *supra* note 30, at 323–327 (on eliminating mens rea for all crimes). See generally N. Kittrie, The Right to be Different (1971).

33. These factors overlap. Note, moreover, that the operational difference between the punishment and therapeutic models is not always clear. See, e.g., Shapiro & Spece, *supra* note 1, at 171–174. Some therapeutic models may be disorder-oriented; others may favor a behaviorist notion of maladaptive behavior. For a review of the idea of models in general and of various models of intervention, see *id.* at 158–174.

34. But recall that on some views one's character (a status of sorts) is always potentially in issue. See *supra* note 25. One who can establish that he or she acted inconsistently with his or her character because of external compulsion or internal disorder has at least the beginnings of a case for mitigation or exculpation. Note that this reflects a view that the suspension of the usual causes is a reason for questioning culpability. See Wolf, *supra* note 29. On the requirement of a culpable act, see generally Kadish & Schulhofer, *supra* note 30, at 188–217. See also Robinson v. California, 370 U.S. 660 (1962) (violation of Eighth Amendment to punish for status of being a narcotic addict). The inchoate offenses (e.g., attempt, conspiracy) require conduct, not just inclinations. See G. Fletcher, Rethinking Criminal Law 131–135, 218–223 (1978).

35. But note previous attempts to predict behavior without the use of hard physiological models. For a discussion of various efforts to assess dangerousness, see C. Slobogin, *Dangerousness and Expertise*, 133 U. Pa. L. Rev. 97 (1984).

36. This of course goes far beyond Tarasoff v. Regents of the University of California, 17 Cal. 3d 425 (1976) (cause of action recognized against therapist—in this case a psychologist—for failure to warn victim or others who were likely to advise her of patient's intention to kill her).

37. Some observers criticize the civil commitment system as a form of prevention detention. S. Morse, *Crazy Behavior, Morals, and Science: An Analysis of Mental Health Law*, 51 S. Cal. Rev. 527, 631–632 (1978) ("Commitment of only the mentally ill is an unfair and unsuccessful means for achieving the primary goals of preventive detention . . ."). And some may see drug or AIDS screening programs as nascent moves in that direction.

For an analysis of screening for violent propensities, see P. Brown, *Guilt by Physiology: The Constitutionality of Tests to Determine Predisposition to Violent Behavior*, 48 S. Cal. L. Rev. 489 (1974). See also M. Shapiro & R. Spece, 1991 Supplement to Bioethics and Law: Cases, Materials and Problems 69 (the Supreme Court and future dangerousness). See generally E. Richards, *The Jurisprudence of Prevention: The Right of Societal Self-Defense Against Dangerous Individuals*, 16 Hastings Const. L.Q. 329 (1989).

38. "By bringing us face to face with such questions [as instant voting by the entire electorate on various issues], Dr. Taviss argues, technology has the effect of calling society's bluff and thereby preparing the ground for changes in its values. . . . Technology . . . has the effect of facing us with contradictions in our own value system and of calling for deliberate attention to their resolution" (E. Mesthene, Technological Change: Its Impact on Man and Society 53 [1970]).

39. See *infra* note 12.

40. See Snyder, *supra* note 1. Other organic therapies have also been investigated, e.g., electroconvulsive therapy (thought by many to be very effective, but still controversial) and

psychosurgery (some data on effectiveness, but very controversial). See Shapiro & Spece, *supra* note 1, at Part II. See also Potter et al., *supra* note 1, reviewing antidepressant therapy.

41. 494 U.S. 210 (1990). For a discussion of the use of behavior control technologies justified under models not focusing on disorder (e.g., custodial control), see Shapiro & Spece, *supra* note 1, at 158–165, 179–203, 210–224.

42. But cf. Riggins v. Nevada, 112 S. Ct. 1810, 1815 (1992) (involuntary administration of antipsychotic drug during criminal trial violated defendant's Sixth and Fourteenth Amendment rights; under Washington v. Harper, "overriding justification and a determination of medical appropriateness" is required for such compulsion). Presumably the idea of a liberty interest also applies to other psychotropic drugs, though different sets of effects may alter the Court's description of the relevant liberty interests and of the state's interests in administering the drugs. The Court has not expressly considered the right-against-treatment issue as applied to persons confined in mental health facilities. In an earlier case, Mills v. Rogers, 457 U.S. 291 (1982), the Court assumed for argument's sake that involuntarily committed mental patients have a liberty interest in avoiding unwanted administration of antipsychotic drugs but did not decide the case, remanding it because it may have been decided under state rather than federal law.

Note that in Harper, the Court adopted a standard of review far short of "strict scrutiny," strongly deferring to the judgment of medical and prison administration professionals. Cf. Youngberg v. Romeo, 457 U.S. 307 (1982) (relied on in Harper), adopting a standard in which courts defer to professional judgment in deciding whether a liberty interest in safety and freedom from restraint can be overridden. The presumption favoring professional judgment can be overcome on a showing of a departure from accepted professional judgment so substantial as to demonstrate that there was no professional judgment at all. But cf. Riggins v. Nevada, *supra* (requirement of overriding justification). The Court seems to be well into the process of editing the ideas of fundamental rights and strict standards of review out of certain areas of constitutional adjudication. See, e.g., Cruzan v. Director, Missouri Department of Health, 497 U.S. 261 (1990) (apparently accepting the idea of a liberty interest in avoiding unwanted medical treatment, including some forms of lifesaving treatment; no "fundamental rights" or "privacy rights" found; broad scope accorded to state to pursue its interests in ways that invade the liberty interest); Webster v. Reproductive Health Services, 492 U.S. 490 (1989) (plurality opinion expressed preference for referring to abortion rights as "liberty interests" rather than "fundamental rights" and declined to apply notions of "privacy"; greater scope for state regulation recognized under a standard of review not fully specified). But cf. Foucha v. Louisiana, 112 S. Ct. 1780, 1785 (1992) (referring to the "fundamental nature" of the right to liberty in the sense of freedom from bodily restraint). And see Michael H. v. Gerald D., 491 U.S. 110 (1989) for a Supreme Court debate on the recognition of liberty interests. State courts may, of course, accord more expansive rights under their own laws. See generally Shapiro & Spece, *supra* note 37, at 55–57.

43. Compare the use of Depo-Provera to reduce sexual aggressiveness in males. See generally E. Fitzgerald, *Chemical Castration: MPA Treatment of the Sexual Offender*, 18 Am. J. Crim. L. 1 (1990), and Shapiro & Spece, *supra* note 37, at 47–52. But it is not a magic bullet, either. See Fitzgerald, *supra*, at 7–10 (discussing side effects of Depo-Provera and lack of data concerning its long-term effect on recidivism).

44. See Shapiro & Spece, *supra* note 1, at ch. 6 (behavior control in prisons and mental hospitals).

45. I am not suggesting that without modern technology effective behavior control is impossible. Behavior can be shaped by education and other influences, as well as by coercion, incentives, and so on.

46. Note the theoretical possibility of germ-line augmentation. For complex traits such as intelligence, however, the process would be very difficult. See Shapiro & Spece, *supra* note 37, at 97–101.

47. But such augmentation might be applied within a larger prevention-of-harm model that posits reduced risks resulting from enhanced intellect.

48. See generally R. Delgado, *Organically Induced Behavioral Change in Correctional Institutions: Release Decisions and the "New Man" Phenomenon*, 50 S. Cal. L. Rev. 215 (1977).

49. See generally G. Kolata, *Genetic Screening Raises Questions for Employers and Insurers*, 232 Science 317 (1986). On the genome mapping project, see J. Watson & P. Cook-Deegan, *The Human Genome Project and International Health*, 263 J.A.M.A. 3322 (1990); Shapiro & Spece, *supra* note 37, at 118–121. The possibility of predicting behavior or the occurrence of disorder from particular sets of genome information may be limited, however.

50. See generally L. Tribe, American Constitutional Law 1329–1335 (2d ed. 1988).

51. Cf. Whalen v. Roe, 429 U.S. 589 (1977) (tentative acknowledgment of a constitutional interest in informational privacy, but holding that under the circumstances, a computerized record of prescriptions of restricted drugs did not threaten it). See also Winston v. Lee, 470 U.S. 753 (1985) (compelled surgery to obtain evidence from an accused's body was unreasonable under the Fourth Amendment, given the circumstances). But cf. also Schmerber v. California, 384 U.S. 757 (1966) (extracting blood from person involved in auto accident to determine alcohol content was a reasonable search under the Fourth and Fourteenth Amendments; no violation of privilege against self-incrimination or of due process); Calif. Health & Safety Code sec. 199.21 (confidentiality of AIDS test results). On genetic screening issues, see generally M. Rothstein, Medical Screening and the Employee Health Cost Crisis (1989).

52. Depending on the circumstances, one might also raise Fourth and Fifth Amendment claims. See Brown, *supra* note 37, at 511–527. See generally D. Merritt, *Communicable Disease and Constitutional Law: Controlling AIDS*, 61 N.Y.U. L. Rev. 739 (1986).

53. Moore v. Regents of University of California, 51 Cal. 3d 120, 133, 271 Cal. Rptr. 146 (1990). See generally Office of Technology Assessment, New Developments in Biotechnology: Ownership of Human Tissues and Cells (1987). See also M. Havens, *The Spleen That Fought Back*, 20 The Brief 11 (1990) (commenting on the Moore case and endorsing recognition of privacy rights against commercial exploitation of one's biological parts).

54. See generally H. Hansmann, *The Economics and Ethics of Markets for Human Organs*, 14 J. Health Pol. Pol'y & Law 57, 76–77 (1989), noting shifts in normative categories over time in response to experience with innovations—e.g., revision of our views on artificial insemination. He observes that "[t]ransactions can be and have been recategorized when technological changes have made market mechanisms advantageous." *Id.* at 76–77. Note that some of these reconstructions may be occasioned by "fragmentation" of various natural and human processes. See section IV, *infra*.

55. This is R. Carnap's term, from Logical Foundations of Probability 576 (2d ed. 1962). I do not claim that its present use is what he had in mind. See generally the discussion in Shapiro & Spece, *supra* note 1, at 18–22.

56. H.L.A. Hart noted this possibility in *Positivism and the Separation of Law and Morals*, 71 Harv. L. Rev. 593, 622 (1958):

The world in which we live, and we who live in it, may one day change in many different ways; and if this change were radical enough not only would certain statements of fact now true be false and vice versa, but whole ways of thinking and talking which constitute our present conceptual apparatus, through which we see the world and each other, would lapse. We have only to consider how the whole of our social, moral, and legal life, as we understand it now, depends on the contingent fact that though our bodies do change in shape, size, and other physical properties they do not do this so drastically nor with such quicksilver rapidity and irregularity that we cannot identify each other as the same persistent individual over considerable spans of time. Though this is but a contingent fact which may one day be different, on it at present rest huge structures of our thought and principles of action and social life.

Gradual changes over an extended period are generally viewed differently, at least when arising from "normal" courses of growth and development. See E. Shils, *The Sanctity of Life*, in Life or Death: Ethics and Options 2, 32–36 (D. Labby ed., 1968). Nevertheless, there are no clear criteria specifying the limits of appropriate human transformation.

57. For a more elaborate tracing of the consequences of the power to "divide and conquer" life processes, see M. Shapiro, *Fragmenting and Reassembling the World: Of Flying Squirrels, Augmented Persons, and Other Monsters*, 51 Ohio St. L.J. 331 (1990). See also N. Davis, *Reproductive Technologies and Our Attitudes Towards Children*, 9 Logos 51 (1988). There are of course many "fragmentations" that do not necessarily pose challenges to existing descriptive and normative categories—e.g., an appendectomy.

58. E.g., Cal. Health & Safety Code sec. 7180 (West Supp. 1992).

59. See Cruzan v. Director, Missouri Department of Health, *supra* note 42, particularly Justice Stevens's dissenting opinion.

60. See, e.g., M. Green & D. Wikler, *Brain Death and Personal Identity*, 9 Philos. & Pub. Aff. 105 (Winter 1980). See generally the discussion and references in Shapiro & Spece, *supra* note 37, at 202–204; D. Lamb, Death, Brain Death and Ethics (1985).

61. See generally M. Shapiro, *The Technology of Perfection: Performance Enhancement and the Control of Attributes*, 65 S. Cal. L. Rev. 11, 56–57 (1991); M. Shapiro, *Who Merits Merit? Problems in Distributive Justice and Utility Posed by the New Biology*, 48 S. Cal. L. Rev. 318 (1974).

62. See Shapiro & Spece, *supra* note 37, at 127–132. Compare the use of anencephalic infants as organ donors (in those cases, of course, conception was not aimed at transplantation). See Anencephalic Organ Donation Committee of Loma Linda University Medical Center, *Considerations of Anencephalic Infants as Organ Donors—A Working Document*, 12–18–87 (protocol); J. Friedman, *Taking the Camel by the Nose: The Anencephalic as a Source for Pediatric Organ Transplants*, 90 Colum. L. Rev. 917 (1990). Note also the problem of deciding to have a child in the hope that it can serve as a tissue donor. See Shapiro & Spece, *supra* note 37, at 314–317 (child of the Ayalas provided bone marrow for transplantation into older sister).

63. D. Wiggins, *Deliberation and Practical Reason*, in Essays on Aristotle's Ethics 221, 233 (A. Rorty ed., 1980). He says, preceding the text quotation: "No theory, if it is to recapitulate or reconstruct practical reasoning even as well as mathematical logic recapitulates or reconstructs the actual experience of conducting or exploring deductive argument, can treat the concerns which an agent brings to any situation as forming a closed, complete, consistent system. For it is of the essence of these concerns to make competing and inconsistent claims. (This is a mark not of irrationality but of *rationality* in the face of the plurality of ends and the plurality of human goods.) The weight of the claims represented by these concerns is not necessarily fixed in advance. Nor need the concerns be hierarchically ordered" (emphasis in original).

I am not suggesting that changes of circumstances automatically produce massive changes in our conceptual systems. On "cognitive inertia," see Margolis, *supra* note 12, at 183 (strong efforts required to alter deeply entrenched habits of mind); Shapiro, *supra* note 57, at 353–354 (raising doubts about "commodification" and "objectification" processes in certain contexts); S. Altman, *(Com)modifying Experience*, 65 S. Cal. L. Rev. 293, 325–329 (1991).

64. These might be considered to be a part of the conceptual and normative reevaluation discussed in the previous subsection.

65. B. Mazlish, *The Fourth Discontinuity*, Technology and Culture 1 (Jan. 1967). A brief reference to the mind-body problem is in order here. The idea that propositions about mental states or processes are "reducible" to propositions framed in physicalist terms is not logically connected to the breakdown of continuities discussed by Mazlish. Nevertheless, the communication of reductionist views might accelerate the psychological erosion processes that Mazlish describes.

66. See subsection (IV-A-4), *infra.*

67. Cf. M. Heidegger, Being and Time 102–03 (J. Macquarrie & E. Robinson trans., 1962): "When its unusability is . . . discovered, equipment becomes conspicuous. This *conspicuousness* presents the ready-to-hand equipment as in a certain un-readiness-to-hand" (emphasis in original).

68. See generally G. Winslow, Triage (1982).

69. See the reference in note 35, *supra.*

70. See Shapiro & Spece, *supra* note 1, at 17–18.

71. Some argue that certain decisions by persons in or threatened with captivity cannot be voluntary. See (describing and criticizing that view) J. Murphy, *Total Institutions and the Possibility of Consent to Organic Therapies*, 5 Human Rights 25 (1975).

72. There is an instructive literature on how various groups deal with internal inconsistencies and gaps in their belief systems. See, e.g, B. Myerhoff & S. Moore, Symbol and Politics in Communal Ideology (1975); B. Myerhoff & S. Moore, Secular Ritual (1977). See generally G. Calabresi & P. Bobbitt, Tragic Choices (1978).

73. There is a serious problem about "intertranslatibility" in judging one conceptual system from the perspective of another. There is no reason to pursue this point here. See A. MacIntyre, Whose Justice? Which Rationality? 370–388 (1988), for a discussion of translatability, incompatibility, and incommensurability.

74. See the references suggested in note 63, *supra,* on cognitive inertia.

75. And perhaps not. Some may argue that technological augmentation of our performance capacities do not separate us from things and objects—they confirm the connection and do not improve us as persons. They merely provide us with unearned benefits. See the discussion of the "paradox of perfectionism" in Shapiro, *supra* note 57, at 355–356, and Shapiro, *supra* note 61, at 34–36.

76. See Shapiro, *supra* note 57, at 363–365.

77. See generally Shapiro, *supra* note 57, at 348–357. The formal and informal regulatory systems that implement the new technologies also may shape attitudes and beliefs simply by being observed. See M. Shapiro, *On Not Watering All the Flowers: Regulatory Theory and the Funding of Heart Transplantation*, 28 Jurimetrics J. 21, 23–27 (1987).

78. See generally G. Dworkin, *Is More Choice Better Than Less?* 7 Midwest Studies in Philos. 47 (1982); Shapiro, *supra* note 57, at 349–350. See also D. Callahan, *Science: Limits and Prohibitions*, 3 Hastings Center Rep. 5, 6 (Nov. 1973) (new options derived from technology may eventually lead to limits on options).

79. But see the National Organ Transplant Act of 1984, forbidding the transfer for valuable consideration of organs to be used in transplantation if such transfer affects interstate commerce, 42 USCA sec. 274e West 1991.

80. P. Greenspan, *Free Will and the Genome Project*, 22 Philos. & Pub. Aff. 31 (1993), which was published while this article was in press, presents arguments paralleling some of those offered in the preceding sections.

Part 5

Implications for Law and Policy

Given the complexity of our legal system, how might knowledge of the biochemistry of human social behavior best be integrated into contemporary law and public policy?

13

The Lesson of the Owl and the Crows

William H. Rodgers, Jr.

Abstract: The law is a complex system that evolves in terms of its own history and structure. Strategies of deception and counterdeception are used by legislators, judges, and lawyers much as they are used by predator and prey species. As a result, as the field of environmental law illustrates, attempts to make legal concepts more "realistic" often have paradoxical results.

I. GAME THEORY AND THE ADVANTAGES OF THE GOOD FAKE

In recent times, a number of writers have devoted attention to the application of game theory to legal phenomena.[1] Unlike much formal decision analysis, the game theory "best" choice is strikingly conditional and dependent upon the strategies of the other players.[2] The point is made convincingly by the predicament of the red squirrel.[3] The squirrel developed chattering as a strategy to discourage predators like the great horned owl; it was a good strategy for the moment because an alert prey is of no interest to the owl. But chattering is not a good strategy for all times and occasions; with the appearance of a new predator—humans armed with rifles—chattering is a decidedly poor strategy.

In this world of strategies and counterstrategies, the advantages of the good fake are not to be overlooked. Fakery is an indelible part of the landscape in settings where we readily accept the gaming metaphor—sporting events are the obvious examples. But I wish to emphasize how fakery and deception can play an important

This chapter was originally delivered as a lecture at Florida State University College of Law, February 25, 1988. It benefited from an earlier presentation at a workshop at the Yale Law School, November 1986. A companion article appeared as "The Lesson of the Red Squirrel: Consensus and Betrayal in the Environmental Statutes," 5 *J. of Contemp. Health Law & Pol.* (1989) (Catholic University School of Law).

role in legal interactions as well, particularly in the writing of the environmental statutes.

Environmental lawyers often are fond of borrowing examples from natural history to illustrate propositions of law. There is more to this practice than habit, it seems to me, because the natural laws of evolution contain lessons in the results of competitive struggle that fit closely the experiences of legal game playing. Nature, it turns out, offers a treasure trove of fakes, feints, and deceptions, described in the literature usually under the heading of mimicry.[4] Individual members of species succeed not only by hiding but also by misrepresentation if they are discovered by predators. One can find examples of sheep in wolves' clothing: delicious insects that look like poisonous ones, flies that succeed in droning on the same frequency as stinging bees. Some wolves are dressed in sheeps' clothing: one tropical hawk drifts about like a vulture and enjoys great hunting success because the victims mistakenly believe they are in the presence of a bird who prefers carrion. Nature offers the widespread resort to deceptive props: some moths have evolved eye-spot patterns on their wings that are thought to startle predators and thus offer the intended victims momentary opportunities for escape. And nature gives examples of deliberately deceptive behavior: the title of my chapter is drawn from a tale[5] of an owl and some crows who were reared in close proximity to one another. In order to discourage a steady and dangerous diet of silent swoops (any one of which could be fatal), the crows developed a strategy of wandering into easy range "pretending" to be wholly unaware of the presence of the owl, only to sidestep the futile strikes with disdain and ease. By this stratagem the crows "proved" to the satisfaction of the owl that crows could not be seized, under even the best of circumstances, and the unwanted attacks ceased altogether.

II. FAKING AND DECEPTION IN THE DEVELOPMENT OF THE ENVIRONMENTAL STATUTES

In another paper,[6] I have developed the notion that the process of enactment of the environmental statutes is a game-playing phenomenon that is likely to yield legislation with both "consensus" and "betrayal" features. It is important to underscore the significance of deception in building both the "consensus" necessary for enactment and in accomplishing the "betrayal" often observed in the waning hours of enactment.

A. Fakery: The Material of Consensus

1. Process Entitlements, Hedges, and Bets

It is no secret that process can consume substance,[7] and the environmental laws offer many examples.[8] Process is an excellent hiding ground for lawmakers who wish to remain uncommitted to particular substantive outcomes. Students of the subject often are astounded by the profound indirection that grips the field of environmental law. Only occasionally does the law ask whether a particular course of conduct is compatible with environmental values. The question is more likely to be posed in process terms—expressed most prominently in the impact statement requirements.

2. Ambiguity and Delegation

Delegation is another famous preference-hider that makes a regular appearance in the environmental statutes. To mention but one prominent example, the term "unreasonable risk" appears 38 times in the Toxic Substances Control Act (a statute of 64 pages).[9] There is great comfort for legislators who wish to hide in this thicket immune from the sharp scrutiny of predatory constituents.

3. Dissembling and Manipulation

These two techniques are other devices for having it both ways that make regular appearances in the environmental statutes. Dissembling should be taken to mean "hoping to achieve A by voting for B and manipulation means a packing of the agenda to make dissembling possible."[10] The phenomenon is described convincingly by Judge Abner Mikva who gives this account of the "careful nurturing" of a "difficult coalition" (miners, mine owners, environmentalists) by Rep. Morris Udall as he guided the strip-mining legislation through Congress:[11]

> He was the floor manager and a lot of us were sitting in the cloakroom when one of the Members from West Virginia, a mining state, got up and said, "Now will the gentleman from Arizona assure me that this bill protects state sovereignty and makes clear that the states continue to have an active and important role in how the strip mines are to be regulated." Udall said, "The gentleman from West Virginia is absolutely correct. The bill protects states' rights and state sovereignty and makes sure that the states continue to play an important role." A little later on in the debate one of the environmentalist Congressmen got up and said, "Now will the gentleman from Arizona assure me that this once and for all sets federal standards and makes it clear that there is a federal law that decides what kind of strip mining will be allowed?" And Udall said, "The gentleman is absolutely correct. The law once and for all will put the federal authorities in control." He then came into the cloakroom for a drink of water and we laughed and said, "Mo, they both can't be right." He said, "The gentleman is absolutely correct." The bill passed. Is it any wonder that the bill is not as clear and precise and specific as some of us might have wanted?

4. Postponements

Postponements or "more study" provisions are another useful prop for lawmakers who will not or cannot decide. There are all kinds of reasons for putting off a decision until tomorrow, both benign and otherwise, so it is relatively easy for a lawmaker to justify a "more study" vote on principled grounds.

5. Self-Nullifications

These are legislative standoffs "where command and countermand are stuffed into the same sorry package."[12] Good examples are the whistleblower protection provisions that are found in some of the environmental statutes.[13] These offer protection for workers who give evidence of polluting behavior but with a 30-day statute of limitations that withdraws a meaningful remedy in all but the unusual case. Nobody wins under this kind of legislation, of course, but nobody loses either, and that is why self-nullification in legislation is not at all uncommon.

6. Teases or Aspirational Commands

These are additional techniques that allow lawmakers to be simultaneously "for" and "against" identifiable outcomes. Like the delegation tactic, commitment to a vague goal does not tie a lawmaker to any particular means of implementation.[14] This is a dream political world without unsavory options and irascible victims. The hard choices must await initiatives by somebody else.

B. Fakery: The Material of Betrayal

1. Defections

A defection is a term I have used to describe various end-of-the-game stratagems to seize advantages at the expense of the unprepared and ill-equipped. An example is the legislative decision that allowed the Tellico Dam to go forward in the face of powerful environmental and economic objections: "The pork-barrel proponents, in forty-two seconds, in an empty House chamber, were able to slip a rider onto an appropriations bill, repealing all protective laws as they applied to Tellico and ordering the reservoir's completion. Despite a half-hearted veto threat by President Carter and a last-minute constitutionally-based lawsuit brought by the Cherokee Indians, the TVA was ultimately able to finish the dam, close the gates, and flood the valley on November 28, 1979."[15]

2. Sleepers

Sleepers, by definition, are legislative provisions with practical consequences far outstripping those anticipated by the formal legislative vision. It goes without saying that gameplayers who create or detect beneficial "sleepers" do not advertise the fact since the strategy is to go for sizeable gains without paying any price in return. Obviously, pursuit of the "sleeper" strategy means that a law will look a great deal different the day after enactment than it does the day before.

III. THE CONSEQUENCES OF FAKERY OVER TIME: AN EVOLUTIONARY VIEW OF STATUTORY INSTRUCTIONS

A. Instability of the Legislative Product

Can we say anything useful about the fate of these environmental statutes that include a precarious mix of consensus and defection traits? First, let us define a statute as an act of the legislature that instructs persons and organizations to act in certain ways. The typical statute contains many instructions or norms within the legislative package, and these are subject to differing interpretations. This universe of plausible interpretations is taken to represent the population for purposes of analysis. Students of evolution emphasize that close attention must be paid to the sources of change within the population under study and to the selection mechanisms that yield differential survival rates.[16]

What are the mechanisms that can introduce change (or variation) into the instructions turned loose as a statute? Preliminarily, it is intuitively plausible that future game players will take a strongly instrumental view of statutory instructions. These instructions will be treated as part of the background environment available for self-maintenance against the mischief of the world. The statute suddenly means what lawyers (and others) say it means as it undergoes translation in the world at large. Put somewhat differently, statutory instructions immediately will be put to use (as interpretations) in legal game playing in a host of different arenas.

After enactment, self-imposed constraints on the players are suddenly loosed. Players are free to urge interpretations heretofore undisclosed, or indeed unimagined. Sleepers, skewers, defections are now in the open claiming unsuspecting victims. This frenzy of interpretation in the wake of enactment produces a long list of surprising readings and tactics. These include betrayals of sidebar agreements not recorded in the legislation, the invention of legislative history after-the-fact, the attribution of meanings scorned by the negotiating principals.[17] Legislative insiders routinely are offended by the extraordinary directions "their" legislation takes at the hands of outsiders, amateurs, and skewees.

This strong centrifugal tendency towards different interpretations is reinforced by catalysts and accelerators that are part of the initial consensus product. The opportunistic decision-avoidance techniques necessary to secure a nonzero sum law[18] feed these later changes: the postponements, such as the study or commission provisions, yield an empirical harvest that can be used to modify tentative legislative conclusions. Delegations are famous for returning to haunt those who took the easy way out. Other process deferrals rebound in similar fashion as experience proves them unworkable in fact.[19] Variances and exemptions necessary for enactment look like ugly "gaps" from the fuller vantage point offered by time. Teases and half-laws are built-in invitations to complete the story.[20] The victims of 11th-hour defections are spoiling to settle the score and may have good reason to believe they were done in by a bogus or unrepresentative law.[21]

B. Courts as Selection Mechanisms

In this blizzard of different meanings, it is hardly surprising that some interpretations stick, and some do not. The institutions that do the selecting are no mystery (legislatures, courts, agencies, private organizations), and the criteria for selection are the subject of unceasing argument. Our attention will be confined to the courts that have the prerogative to select and annoint as authoritative only the chosen interpretations.

Courts change statutes (as defined in this paper) because they reject some interpretations as implausible and unjustified. Judicial readings can work in tandem with other influences to encourage legislative amendments. The environmental statutes are amended regularly, often with an expressed design to embrace[22] or overrule[23] provocative judicial decisions. A few of the ways in which courts can accelerate this process of legislative reconsideration include the following:

•giving meaning to vague and ambiguous terms that were used as refuges for choice avoidance;[24]

•disregarding contrived legislative history;[25]

•recognizing and enforcing "sleepers" that may change costs and benefits as calculated by the legislature;[26]

•disregarding the cross-cancellation purposes hidden in statutory "teases," which is another way of creating a constituency aggrieved by the turn of events.[27]

It goes without saying that identification of the "correct" posture of the courts in reviewing statutory hybrids (partly consensus, partly betrayals) is no easy task.[28] Game theory counsels that undeviating strategies are doomed to failure. This was the problem of the red squirrel, and it was the problem that was overcome by the crows. One possibility is that the "best" strategy for the courts is neither deference nor a lack thereof to the legislative product, but rather a combination of the two that can resist exploitation by the legislative branch.

IV. EVOLUTION AND DIRECTION: RISE, FALL, OR DRIFT?

History is littered with the reputations of scholars who have struggled to discover normative advice in evolutionary behavior.[29] But the question is unceasingly seductive: do statutes improve over time,[30] decline, or wander about pulled to and fro by the warring factions? This question does not get much easier if an attempt is made to reformulate it in descriptive terms: does the legal presence of statutes (e.g., their authoritativeness or influence) grow, decline, or remain static when measured over time?

A. Problems of Definition and Sampling

It is no easy trick to characterize the viability or authority of any set of legislative instructions. One can envisage any number of indicators (pages, citations in lawyers' documents, recognition in the relevant culture, probability estimates of compliance) that might serve as a rough measure of resolving power or authoritativeness of a legislative directive. Measuring change over time of this elusive "authoritativeness" is a more difficult challenge yet. We have complicated the task still further by defining statutes as accommodating change without the necessity of formal amendment. Impossibility theorems are likely to stand in the way of attempts to evaluate a set of statutory instructions and say something useful about their future prospects.[31] Even so simple a matter as statutory extinction (marked by a repeal) may be indecisive if the observer chooses to pay attention to the genetic heritage of the new legislative species.[32]

Even apart from the standard of statutory "authority," there is the further complication of the short-run sample affording a misleading indicator of trends.[33] In all likelihood, empirical substantiation is available for the more popular models—gradual decline,[34] dormancy and explosive growth, cyclical growth and decline, rapid growth and gradual decline. This is not surprising since the phenomenon under investigation (evolution of populations over time) is dramatically conditional: extravagant success and abject failure alike are subject to sudden reversal when the environment changes. The red squirrel hypothesis advises that every successful strategy is uniquely dependent upon what the other players are doing.[35]

B. The Prospects of Devolution

Finessing these several difficulties, do we nonetheless have reasons for believing that the prominent fate of a statute is decline? After all, information theory is intimately linked to the entropy laws,[36] and legal researchers are constantly stumbling across the bones of old and forgotten laws. Unfortunately, any general thesis of decline also must account for spectacular episodes of growth. Historians of environmental law could substantiate the claim that dramatic "advances" in the field have been occasioned by the opportunistic application of old and outdated laws.[37] (These unforeseen applications of statutory "sleepers" typically inspire "corrective" legislative responses.)

Three lines of argument hold promise for substantiating a thesis of general decline.

First is the intuitively appealing notion that any legislative "snapshot in time," however cleverly contrived, will be isolated and gradually overwhelmed by changes of fact and value. Working against this vision of statutory entropy are various rehabilitation mechanisms to keep the laws "current" or "responsive." Judicial interpretation is an obvious possibility,[38] although one suspects there is a penalty (measured by a decline in our unidentified standard of statutory influence; call it a cost of political illegitimacy) associated with the best judicial revisions of old statutes. Despite this sense of a "Second Law" at work (the thing will run down faster than it can be fixed), we really cannot be sure. The legislation perceived as a "snapshot in time" also evolves with the times. The outcome of any contest between the tendencies of decline and the vehicles of rehabilitation appears indeterminate.

A second way to think about statutes declining over time focuses upon the psychology of the players. Statutory paradigms in the minds of lawyers may share a fate similar to scientific paradigms in the minds of scientists.[39] Old visions are squeezed out by the new not because of mass conversion but because the old believers die and pass from the scene. (An informal tally of subscribers to the old Refuse Act as a tool of water pollution control discloses a diminishing pool of supporters, made up mostly of aged sentimentalists and a small group of the legally illiterate [nonlawyer investigators who naively believe that the law means what it says].) Of course, this metaphor is imperfect: new scientific theories catch on and spread presumably because they represent an improvement in explanatory power and utility for the investigator. Are new laws an improvement on the old in this practice-enhancing sense? (Old laws, after all, have stood the test of time.) One can think of a half-dozen reasons why "new law" arguments are preferable to the "old" (e.g., greater claim to legitimacy, closer fit to contemporary problems, more efficient according to 9 of 10 lawyers). But strong trends do not necessarily exclude opportunities for the maverick. Game theory brings us back to the lesson of the red squirrel and of the owl and the crows: every strategy is a winner if the others are pursuing strategies that make it a winner.[40]

A third way to assess the fate of statutes over time can draw upon the insights of evolutionary biology about the dangers of specialization. Exquisite refinement within clumsy historical constraint is an evolutionary path oft taken,[41] and it is a road to extinction: "Such is the way of nature that solutions so perfect they left no room for further improvement often [come] to a premature end."[42] This course of specialization is followed by many of the environmental statutes for reasons having to do with the dynamics of the game playing that produce the legislation and changes to it.[43] One can expect these jerry-built specialist laws to be trapped in a destiny of decline, flanked and overwhelmed by circumstances changing about them. The decline of individual enactments, of course, does not speak to the success or failure of the overall enterprise. The legal war against pollution proceeds on many fronts and can claim gains elsewhere that outweigh individual setbacks.

V. CONCLUSION

Deception is an integral component of communication and a prominent feature of the decision making processes that yield the formal products of legislation. The effect of this deception on future legislative behavior and its treatment in the courts are deserving subjects of inquiry. The principal judicial response to this phenomenon has been denial, but there are reasons to believe that consistent perpetuation of this fiction assures neither admirable legislation nor a quick fix of past errors.

NOTES

1. See, e.g., W. H. Rodgers, Jr., *The Evolution of Cooperation in Natural Resources Law: The Drifter/Habitue Distinction*, 38 U. Fla. L. Rev. 195 (1986) [hereinafter cited as 1986 Drifter/Habitue]; W. H. Rodgers, Jr., *The Lesson of the Owl and the Crows: The Role of Deception in the Evolution of the Environmental Statutes*, 4 Fla. State Un. J. of Land Use & Env'tl L. 377 (1989); W.H. Rodgers, Jr., Environmental Law: Air & Water, 2 vols. (1986) [hereinafter cited as 1986 Air & Water].

2. 1986 Drifter/Habitue at 197.

3. W. H. Rodgers, Jr., *The Lesson of the Red Squirrel: Consensus and Betrayal in the Environmental Statutes*, 5 J. Contemp. Health Law & Pol. 161 (1989) [hereinafter cited as 1989 Red Squirrel].

4. E.g., R. Lewin, *Research News: Do Animals Read Minds, Tell Lies?* 238 Science 1350 (Dec. 4, 1987); letter of M.E. Bitterman, *Creative Deception*, 239 Science 1360 (March 18, 1988) (noting that the capacity to communicate includes the ability to deceive); P. Evans, Ourselves and Other Animals, ch.7 (1987).

5. B. Heinrich, One Man's Owl, ch.13 (Princeton U. Press, 1987).

6. 1989 Red Squirrel.

7. G. Hazard, The Effect of the Class Action on the Substantive Law, 58 FRD 307, 307 (1973) ("Substantive law is shaped and articulated by procedural possibilities").

8. E.g., 1986 Air & Water, v.2, § 4.33 at 476–482 (toxic water pollutants).

9. See W. H. Rodgers, Jr., Environmental Law: Pesticides & Toxic Substances, vol. 3, § 6.1 at 372 (1988).

10. 1989 Red Squirrel at 166.

11. *Reading and Writing Statutes*, 48 U. Pitt. L. Rev. 627, 636–637 (1987).

12. 1989 Red Squirrel at 168.

13. See, e.g., Section 23 of the Toxic Substances Control Act, 15 U.S.C.A. § 2622.

14. See D. Schoenbrod, *Goals Statutes or Rules Statutes: The Case of the Clean Air Act*, 30 U.C.L.A. L. Rev. 740 (1983).

15. Z. Plater, *In the Wake of the Snail Darter: An Environmental Law Paradigm and Its Consequences*, 19 U. Mich. J. Law Ref. 805, 813–814 (1986) (footnotes omitted). See *id.* at 813–814 n. 32 (pointing out how the legislative "maneuver" violated the rules of the House of Representatives and how the move was engineered "so that none of the few representatives present would understand what was being done").

16. E.g., N. Eldredge, Time Frames: The Rethinking of Darwinian Evolution and the Theory of Punctuated Equibria (1985); E. Mahr, The Growth of Biological Thought: Diversity, Evolution, and Inheritance (1982). See also J.T. Bonner, The Evolution of Culture in Animals (1980); M. Chibnik, *The Evolution of Cultural Rules*, 37 J. of Anthropol. Res. 256 (1981); R. C. Clark, *The Morphogenesis of Subchapter C: An Essay in Statutory Evolution and Reform*, 87 Yale L.J. 90 (1977); P. Diener, *Quantum Adjustment, Macroevolution, and the Social Field: Some Comments on Evolution and Culture*, 21 Current Anthropol. 423 (1980); H. Kaufman, Time, Chance, and Organizations: Natural Selection in a Perilous Environment (1985).

17. An illustration appears in Lombardo v. Handler, 397 F. Supp. 792 (D.D.C. 1975), aff'd 543 F.2d 1043 (D.C. Cir. 1976).

18. 1986 Air & Water, v.1, preface at vii.

19. The toxic pollutant provisions of the Clean Water Act are illustrative. 1986 Air & Water, v.2, § 4.33. See also C. S. Smith, A Search for Structure: Selected Essays on Science, Art, and History 62, fig. 3.13 (MIT Press, 1981).

20. The "protect and enhance" language in the statement of purposes of the Clean Air Act has grown into a huge no-significant-deterioration program. *Id.*, v.1, §§ 3.21–3.23.

21. Text accompanying note 15, *supra*.

22. 1986 Air & Water, v. 1, preface.

23. W. N. Eskridge, Jr., Overriding Supreme Court Statutory Interpretation Decisions, 101 Yale L. J. 331 (1991).

24. A commendable practice that is disapproved in B. Diver, *Statutory Interpretation in the Administrative State*, 133 U. Pa. L. Rev. 549 (1985).

25. Compare United States v. Ilco, Inc., U.S. Dist. Lexus 6218 (N.D. Ala., July 14, 1987) (disregarding affidavits of two members of Congress stating what was on their minds at the time of enactment in 1980) with A. Mikva, *Reading and Writing Statutes*, 48 U. Pitt. L. Rev. 627 (1987) (on the tendencies to develop contrived legislative history).

26. 1986 Air & Water, v.2, § 4.11 (Refuse Act of 1899).

27. Note 20, *supra*.

28. 1989 Red Squirrel at 170–171, pointing out how the normal assumptions of statutory interpretation disavow the deceptions that are manifest in the process and product.

29. E.g., J. Huxley, Evolution in Action (1953); P. Kitchen, Vaulting Ambition: Sociobiology and the Quest for Human Nature (1985); R. Nisbet, History of the Idea of Progress (1980); G.G. Simpson, The Meaning of Evolution (1966); R.J. Richards, Darwin and the Emergence of Evolutionary Theories of Mind and Behavior (U. Chicago Press, 1987).

30. A poll of all of those who select one interpretation over another is likely to disclose unanimity for the proposition that the selected reading makes the statute "better" (e.g., more welfare enhancing). What is the collective outcome of these countless individual "bests"?

31. See L.B. Slobodkin, Growth & Regulation of Animal Populations 206 (2d enlarged ed. 1980) ("There does not exist any measurement or set of measurements of a population itself which will serve as a measure of its probability of survival").

32. To mention but one example, the Insecticide Act of 1910 was repealed in 1947, but its historic statutory instructions linger on in today's Federal Insecticide, Fungicide, and Rodenticide Act. The search for common ancestry is a delightful exercise long engaged in by paleoanthropologists, among others. It has been aided by the spectacular advances in the techniques of molecular biology.

33. See V. Braitenberg, Vehicles: Experiments in Synthetic Psychology 19, fig. 8 (MIT Press, 1984).

34. See E. R. Tufte, The Visual Display of Quantitative Information 41 (Graphics Press, 1983) (the famous graphic showing the decline in the number of Napoleon's troops in the Russian campaign of 1812–1813).

35. 1989 Red Squirrel. Axelrod shows us that tit-for-tat is a robust and successful strategy in a wide variety of environments. Evolution of Cooperation, passim (1984). See also P. Anthony, Golem in the Gears, ch.16 (1986). Tit-for-tat is a failing strategy, however, in a world of defectors.

36. See J. Campbell, Information, Entropy, Language, and Life (1982). Was it Alfred North Whitehead who said, "Information keeps no better than fish"?

37. Nominees include issues of clearcutting, DDT, water pollution (Refuse Act and Bonneville Act that forbids power deliveries to polluters), and the Alaska pipeline. User (as distinguished from preserver) gains under outdated laws include the Mining Law of 1872 and Massachusetts beach access.

38. G. Calabresi, A Common Law for the Age of Statutes (1982); G. Gilmore, The Ages of American Law (1977).

39. T.S. Kuhn, The Structure of Scientific Revolutions (1962).

40. The discussion in text is suggestive of a famous article by John Maynard Smith, *The Theory of Games and the Evolution of Animal Conflict*, 47 J. Theoret. Biol. 209 (1974). One suspects that the choice of research models is better explained by game theoretic thinking than by maximization analysis (where researchers gradually converge on the "best" approach). In law, it seems, growth of popularity of one school simply opens the door to the successful invasion by "maverick" hypotheses.

41. This lesson is taught elegantly in the several books of Stephen Jay Gould, including The Flamingo's Smile: Reflections in Natural History (1985).

42. L. Margulis & D. Sagan, Microcosmos: Four Billion Years of Microbial Evolution 147 (1986).

43. See note 22, *supra*.

14

Low Serotonin Function and the Therapeutic Power of the Criminal Law

David B. Wexler

Abstract: Law can be understood as a mode of remedial intervention rather than as a purely punitive instrument. This perspective throws valuable light on the implications of research on neurotransmitters and behavior for the criminal law.

In his chapter, Jeffery (this volume) decries the criminal law for its failure to embrace the scientific model of human behavior based on "determinism, treatment, and prevention." And Shapiro's chapter (this volume) rightly tells us to "hold the revolution," because advances in the neural sciences, such as those relating to the serotonin studies, do not inexorably call for the toppling of the criminal law system and its concepts of blameworthiness and responsibility.

As one interested in therapeutic jurisprudence—the study of the role of the law as a therapeutic agent (Wexler & Winick, 1991; Wexler, 1990)—I would like to explore various mechanisms, more or less consistent with existing doctrine, whereby the criminal law may be invoked to further therapeutic and preventive objectives. I will do so in the context of the research of Markku Linnoila and his colleagues (Virkkunen, DeJong, Bartko, Goodwin, & Linnoila; see also chapter 6).

That research relates recidivism to blood glucose and serotonin levels. Specifically, the researchers followed 58 violent offenders and impulsive fire setters for an average of three years following release from prison. During that period, 13 offenders recidivated. From the analysis of fluid samples taken before release, one could correctly classify 84.2% of the subjects as persons who would or would not recidivate within the three year follow-up period.

In terms of treatment and prevention strategies, the authors note that, according to police reports and medical records, the recidivists committed their subsequent offenses "almost without exception while under the influence of alcohol" (Virkkunen et al., 1989, p. 603). They note, too, that because of the offenders' low CSF 5-HIAA concentrations, the offenders may be treatable with serotonergic drugs. Such drugs can ameliorate a serotonergic deficit and have been reported to reduce human alcohol consumption and to improve abnormal glucose metabolism in experimental animals.

If future studies confirm the above findings, it may be reasonable, within existing legal doctrine, to test violent parole-eligible offenders and to ascertain their serotonin and blood glucose levels. This could be done nonconsensually or, if consent were to be required, as a condition of parole consideration; parole might properly be denied to those unwilling to undergo the tests. An appropriate consideration in the determination of "reasonableness," of course, would be the nature and intrusiveness of the testing procedure (e.g., lumbar puncture versus possible alternative approaches).

Moreover, for those whose levels indicate that they will be at high risk for recidivism, a reasonable parole condition could be to abstain from alcohol (or from becoming intoxicated) and, assuming no serious side effects, to regularly take serotonergic medication. Under our existing legal scheme, therefore, the device of reasonable parole conditions can provide appropriate motivation for releasees to embark upon a therapeutic-preventive course.

The parole condition device, however, is commonplace and obvious. But what of persons with behavior control disorders who are no longer under the pressure of parole? Does the criminal law have any therapeutic power with respect to them?

In the first place, as Jeffery notes, the prospect of criminal punishment is supposed to serve a deterrent purpose. Ordinarily, crime prevention through deterrence occurs simply through persons choosing not to violate the law and not to run the risk of incurring criminal sanctions. With regard to biologically impaired impulsive offenders aware of their vulnerability and violence-proneness, the deterrence mechanism would expectedly operate in a somewhat more complicated way. To avoid (or at least minimize) the risk of incurring the criminal sanction, those offenders must do something more than simply vow not to violate the law. They must take serotonergic drugs, avoid alcohol, and so forth. It is of course an empirical question whether they will in fact follow a reasonable therapeutic course in order to avoid a confrontation with the criminal law. If they do, the irony, from Jeffery's perspective, ought to be that the harsher the criminal system, the more it may serve as a motivator for persons to follow the therapeutic, preventive, and scientific path that he chides the law for not providing.

But classical deterrence theory may not work optimally to drive impulse-disordered persons into therapeutic channels. If such persons are not now contemplating the commission of a crime of violence, they may not be motivated to take affirmative steps now to avoid possible future criminal activity that, at the moment, seems to them very remote (cf. Wilson & Herrnstein, 1985, p. 53). For this group of potential offenders, therefore, the ordinary dose of deterrence may need to be augmented by something more. The necessary additional motivational strength may be provided by the laws prohibiting "reckless endangering."

Created in Wisconsin and brought into the mainstream of American jurisprudence through § 211.2 of the American Law Institute's Model Penal Code (Part II, Model

Penal Code and Commentaries, pp. 194–204, 1980), this law subjects to criminal penalty a person who "recklessly engages in conduct which places or may place another person in danger of death or serious bodily injury."[1]

A reckless endangerment provision is now commonplace in state criminal codes. The provision's underlying purpose was well stated by the Arizona Criminal Code Commission (1975, p. 234):

> Section 1200, which penalizes reckless endangerment, is necessary to supplement the law of criminal attempt. Under the law of attempt . . . , an individual who intends to cause the death of another but fails to achieve his or her objective can be punished for attempted murder. A person who intends to inflict bodily injury upon another but fails can be punished for attempted assault. But a person who recklessly engages in conduct likely to cause the same results but fortuitously fails to do so cannot be convicted of criminal attempt because of the requirement . . . of a specific intent to commit a crime. In other words, a person cannot intend to act "recklessly" or with "criminal negligence" toward a result and therefore cannot commit criminal attempt for a crime having one of these types of culpability as its essential mental state. As with attempt, § 1200 reflects that certain conduct short of causing physical injury or another offense of greater dangerousness is sufficiently dangerous to physical well-being to warrant imposition of criminal sanctions.

Both the letter and the spirit of reckless endangerment laws seem to support their application to persons with medical conditions who fail to behave responsibly and accordingly pose a substantial risk of serious injury to others. Not surprisingly, the paradigmatic "medical" case involves the AIDS virus. In *People v. Hawkrigg* (1988), for example, a reckless endangerment prosecution was brought where "the defendant, knowing he had AIDS, engaged in deviate sexual intercourse with a youth when he also knew such conduct was a means of transmission of the AIDS virus" (p. 754).

In many jurisdictions, reckless endangerment may be a felony or a misdemeanor, depending upon the seriousness and imminence of the harm risked. *Hawkrigg* was a New York prosecution for felonious reckless endangerment. One might imagine at least the misdemeanor version being used against a person aware of his or her violence-proneness and aware of the dangers of alcohol consumption who nonetheless fails to follow a reasonable therapeutic-preventive course.

If such a person failed to follow such a course, conscious of his or her heightened risk of violence, and ultimately killed another human being, that individual could quite clearly be prosecuted for some sort of reckless homicide (manslaughter or extreme indifference murder). In fact, he or she could be successfully prosecuted for reckless homicide even if at the time of the fatal incident that person was wholly overcome by his or her disability. For example, a close analogy is provided by *People v. Decina* (1956), where a seizure-prone epileptic who drove a motor vehicle and had a fatal accident was properly convicted of criminally negligent vehicular homicide.

But prosecutions for reckless behavior need not await the ultimate harm. The purpose of reckless endangerment enactments is to criminalize the mere reckless *endangering* of another. In *Hawkrigg*, for example, no proof was required that the virus was actually transmitted to Hawkrigg's partner. And in *People v. Tocco* (1988, pp. 142–143), a felonious reckless endangerment statute was applied to the mere drinking behavior of a chronic alcoholic: "If an alcoholic knows that he is prone to commit criminal acts while drunk and that the consumption of even one drink will destroy his ability to resist further drinking, to the point of intoxication . . . it must follow that his voluntarily imbibing of the first drink is the very initiation of a reckless act —and the concomitant disregard of the substantial and unjustifiable risk attendant thereto. If so, under our law the consumption of said drink by such alcoholic raises the act to the level of recklessness per se. . . ."

With appropriate input from the medical, scientific, therapeutic, and other communities, it is surely foreseeable that a sensible prosecutorial policy might emerge to invoke the law of reckless endangerment against persons aware of their vulnerability to violence who refuse to follow reasonable therapeutic-preventive courses of action.[2] After all, from a rather early age we begin to regard as blameworthy those who fail to control controllable problems of aggression (Sigelman & Begley, 1987; see also Robinson, 1985). And the threat of such enforcement efforts would provide a therapeutic inducement akin to that which exists when an at-risk individual is formally on parole, conditional release, or outpatient-commitment status.[3]

Of course, the prospect of reckless endangerment prosecutions against noncompliers raises the danger of extending the state's net of social control. But that "extension" of the net flows more or less from a creative application of *existing* criminal law and doctrine. And the extension may itself be warranted if the prosecutorial policies are sensibly devised and updated, taking into account scientific information regarding the apparent risk of violence, the imminence and substantiality of such risk, the applicability to this special population of ordinary principles of general deterrence, the viability of therapeutic approaches, the efficacy of serotonergic medication, any possible side effects of the same, the extent to which alcohol consumption increases the risk of violence, and so forth.

In order to mount successfully a prosecution for recklessness, there need to be procedures designed to help the state prove that a noncomplying defendant was aware both of his or her violence-proneness and of the appropriateness of a given therapeutic-preventive course of action. The fact that serotonin-deficient persons are cognitively intact suggests that if they have been effectively warned of the risk they pose and of appropriate measures to avoid it, the prosecution will later be able to establish their conscious awareness and disregard of the risk they pose. Nor would it seem with this population (as might be the case with schizophrenic patients who fail to take their medication) that noncompliance with a suggested therapeutic course would arguably be itself motivated by delusional thoughts or other loss of contact with reality.

The lessons learned above regarding the role of the law in motivating therapeutic pursuits may also teach us something about how we might properly structure a "volitional" insanity test. The traditional *M'Naghten* test of insanity was a purely "cognitive" test, excusing from criminal liability only those whose mental impairments substantially affected their ability to appreciate the wrongfulness of their conduct. The *M'Naghten* doctrine had been widely criticized for many years as being morally underinclusive and psychiatrically crude for its failure to recognize "volitional" impairments, those which might render a person, although cognitively intact, unable to conform his or her conduct to the requirements of the law. Accordingly, the American Law Institute's Model Penal Code (§ 4.01) included both a cognitive and an independent volitional prong in its recommended insanity defense provision. After that action, the volitional (sometimes called "irresistible impulse") test gained markedly in popularity, jurisdiction after jurisdiction augmenting the *M'Naghten* rule with an additional volitional test of exculpation. All of this, however, came to an abrupt end— indeed, to a reversal—after the *Hinckley* verdict. After *Hinckley*, the American law of insanity underwent rapid and massive revision (Wexler, 1985), with the volitional test, among other things, being widely criticized and largely eradicated.

Many of the arguments against the volitional test existed before *Hinckley* but took on new potency after Hinckley's acquittal. The main objections to the test are conceptual (What do we really mean by it? [Fingarette & Hasse, 1979]), adjudicative (How can we distinguish an irresistible impulse from an impulse simply not resisted? This is similar to the conceptual objection but more practical in tone), and utilitarian (Do we really need it? Worse, might it do more harm than good? For example, even if lack of control may more or less in fact exist, might not its legal recognition *itself* have an independent behavioral effect leading certain persons with diagnosed control disorders to even greater law-violating behavior than would otherwise have been the case? Cf. Poirier & Brauner, 1988, p. 5).

Despite the enumerated objections and the rapid decline of a volitional insanity test in American law, there remains a strong feeling that our moral intuitions—and even perhaps our basic constitutional principles (English, 1988)—mandate a volitional aspect of criminal responsibility. And if we regard the insanity defense principally as a jury's exercise of its moral intuitions in the face of a rather complete picture of a defendant's state of mind, cognition, felt pressures, behavioral controls, and so forth, the conceptual and adjudicative difficulties seem less severe, especially if the burden of persuasion is put on the defendant. His or her volitional defense will then succeed only if the defendant can persuade the jury that he or she suffered from a lack of control (compulsive or impulsive, possibly including the serotonin situation) sufficient for the jury to want to excuse him or her.

The real remaining objection to the volitional test, it seems to me, is the practical one that the recognition and application of the test may itself produce additional loss of control (cf. Monahan, 1973; Fein, 1984). To reduce the worry, a volitional test should fail if the defendant culpably caused his or her own lack of control (Morse, 1985, p. 820). If the volitional test is applied in the serotonin context, for example, the jury would learn not only of the research relating serotonin to problems of impulse control and violence, but also of the role of serotonergic drugs, alcohol consumption, and the extent to which the defendant knew of and pursued a reasonable therapeutic course. A "nonculpability" component would make a volitional test more palatable. It would lessen the likelihood that a label of control disorder would, like a self-fulfilling prophecy, lead to additional problems of behavior control by those who are aware of—or those who may even have been prior beneficiaries of—a volitional excuse.

I hope these remarks will enable us to reconcile in practice the theoretically divergent views, set forth earlier in the volume, of Shapiro and Jeffery. Like Shapiro, I have tried not to topple our traditional criminal law concepts merely because of exciting scientific developments. Like Jeffery, I have tried to exploit the therapeutic- preventive aspects of the criminal law. The motivational power of the criminal law can be marshalled to induce dangerous offenders with low serotonin function to take responsibility to control their dangerous disability. Without major alteration of existing legal concepts, the law may nonetheless serve as a therapeutic agent.

NOTES

1. See also Arizona Revised Statutes § 13–1201, which states that "a person commits endangerment by recklessly endangering another person with a substantial risk of imminent death or physical injury"; see also New York Penal Law § 120.20 and § 120.25. A related criminal provision might be that of criminal nuisance, New York Penal Law § 240.45, which renders one guilty of criminal nuisance when "by conduct either unlawful in itself or unreasonable under all the circumstances, he knowingly or recklessly creates or maintains a condition which endangers the safety or health of a considerable number of persons."

2. In the study by Virkkunen et al., serotonin and blood glucose levels enabled the investigators to classify recidivists and nonrecidivists with an accuracy of 84.2%. The accuracy rate for nonrecidivists was 95.5%, and it was 46.2% for the recidivists. While the false positive rate for the recidivists was rather high, the prediction seems good enough to categorize persons in that category as posing a substantial risk of physical harm.

Of course, in many cases, persons who fail to comply with a reasonable therapeutic course of action will not come to the attention of the authorities until some violent action has occurred. In such cases, a prosecutorial policy might specify that those noncompliers be routinely prosecuted for the criminal activity—even for crimes such as simple assault—and that the criminal charge be coupled with a count of reckless endangerment, charging recklessness at the earlier time of therapeutic noncompliance (cf. Robinson, 1985). A prosecutorial strategy of that sort, especially if combined with a judicial willingness to consider consecutive sentences for the two counts, should send a message that compliance with an appropriate therapeutic program is expected and is something society should reasonably expect.

3. Even as to those persons who are reckless through failing to take medication (as opposed to those who are reckless through affirmatively consuming alcohol), a reckless endangerment prosecution is unlikely to encounter serious difficulties regarding the criminalization of "omissions." In most cases, a defendant will actively associate with others in close physical proximity and do others things he ought not to do if he fails to take his medication. In any event, the "omission" issue can be finessed if reckless endangerment prosecutions are based not merely on the failure to take serotonergic medication but also on the affirmative act of alcohol consumption or intoxication—a policy that might make some sense considering the finding of Virkkunen et al. that the serotonin-deficient recidivists in their sample "committed their offenses almost without exception while under the influence of alcohol."

REFERENCES

Arizona Criminal Code Commission. (1975). *Arizona revised criminal code*. Phoenix: Arizona Criminal Code Commission.

English, J. (1988). The light between twilight and dusk: Federal criminal law and the volitional insanity defense. *Hastings Law Journal, 40*, 1–52.

Fein, R. (1984). How the insanity acquittal retards treatment. *Law and Human Behavior, 8*, 283–292.

Fingarette. J., & Hasse, A. (1979). *Mental disabilities and criminal responsibility*. Berkeley: University of California Press.

Monahan, J. (1973). Abolish the insanity defense?—not yet. *Rutgers Law Review, 26*, 719–740.

Morse, S. (1985). Excusing the crazy: The insanity defense reconsidered. *Southern California Law Review, 58*, 777–836.

People v. Decina, 157 N.Y.S.2d 558 (Ct. App. 1956).

People v. Hawkrigg, 525 N.Y.S.2d 752 (Suffolk Cty., 1988).

People v. Tocco, 525 N.Y.S.2d 137 (Bronx Cty., 1988).

Poirier, S., & Brauner, D. (1988, Aug./Sept.). Ethics and the daily language of medical discourse. *Hastings Center Report*, 5–9.

Robinson, P. (1985). Causing the conditions of one's own defense: A study in the limits of theory in criminal law doctrine. *Virginia Law Review, 71*, 1–63.

Sigelman, C., & Begley, N. (1987). The early development of reactions to peers with controllable and uncontrollable problems. *Journal of Pediatric Psychology, 12*, 99–115.

Virkkunen, M., DeJong, J., Bartko, J., Goodwin, F., & Linnoila, M. (1989). Relationship of psychobiological variables to recidivism in violent offenders and impulsive fire setters. *Archives of General Psychiatry, 46*, 600–603.

Wexler, D. (1985). Redefining the insanity problem. *George Washington Law Review, 53*, 528–561.

Wexler, D. (1990). *Therapeutic jurisprudence: The law as a therapeutic agent*. Durham: Carolina Academic Press.

Wexler, D., & Winick, B. (1991). *Essays in Therapeutic Jurisprudence*. Durham: Carolina Academic Press.

Wilson, J.Q., & Herrnstein, R.J. (1985). *Crime and human behavior*. New York: Simon and Schuster.

15

Science Spending and Serotonin

Steven Goldberg

Abstract: Serotonin and its effect on social behavior are an excellent example of the problems that arise in relating scientific research to public-policy priorities. Often, budgetary decisions will be more important than the regulatory process that is generally the focus of discussion.

In analyzing the legal implications of scientific research, there is a tremendous bias toward regulatory issues. Thus, if dramatic correlations are someday found between serotonin levels and certain types of criminal behavior, it is easy to imagine the first wave of policy debate. Should serotonin levels be relevant to the insanity defense? Should mandatory testing be instituted for serotonin levels? Should information about serotonin be provided in the public schools?

But the relationship between law and science does not begin with regulation. In the earlier stages of scientific research, government is deeply involved, particularly because it picks up so many of the bills. Budgetary choices large and small play a vital role in shaping the course of science. The way those choices are made by the government in concert with the scientific community has important implications for the outcome of regulatory issues that lie down the road.[1]

Under the circumstances, it can be useful to look over some unfamiliar terrain in analyzing the legal context of serotonin research. As an example, and because of its importance in its own right, consider the implications of the item veto on research in areas such as serotonin. That is, to be sure, an odd juxtaposition. The item veto—a favorite topic among specialists in constitutional law and budgetary policy—is rarely thought of in terms of emerging areas of scientific research. Yet it is just this sort of odd coupling that is necessary if science policy is to move beyond those regulatory debates that often come after basic issues have been resolved. In particular, as we shall see, the item veto is a mechanism that proponents of serotonin research ignore at their peril.

The United States Constitution grants the president the power to veto bills.[2] Under this language, the president lacks the authority to veto part of a bill while signing the rest into law.[3] By contrast, the constitutions of 43 states give governors the item veto; that is, they have the ability to veto parts of certain bills. Typically, this power is limited to appropriations bills, and thus the item veto is primarily known as a device for saving the taxpayers' money.[4]

In light of the current budget deficit, there have been repeated calls for the institution of the item veto at the federal level. Many governors assert that they are able to balance their state's budgets through use of the item veto.[5] In his 1987 State of the Union address, President Reagan, a longtime supporter of the item veto, said that it would enable him to "carve out boondoggles and pork—those items that would never survive on their own."[6] Supporters of the item veto point to the fact that a president can rarely be expected to veto a huge appropriations bill, complete with nongermane riders, because of his objection to a small number of items. Opponents stress the key congressional role in our constitutional appropriations system and warn against excess presidential power. The issue is not a traditional liberal versus conservative one—supporters of the item veto include Sen. Edward Kennedy while opponents include Sen. Orrin Hatch.[7] President Bush is among the supporters.[8]

Under the circumstances, legal scholars have devoted a fair amount of attention to the item veto. The primary question they have considered is whether Congress could constitutionally enact such a system. Those who believe a statutory item veto would be unconstitutional stress three major points. First, the language of the Constitution, as noted above, gives the president the power to veto bills, not parts of bills. Second, there is some evidence that the framers of the Constitution were aware of the practice of large omnibus appropriations bills and believed that practice was an appropriate exercise of legislative power. Finally, giving the president the power to shape appropriations bills through the item veto might be an improper delegation of Congress's authority in the spending area.[9]

But these objections are hardly dispositive. Supporters of the constitutionality of the item veto note that the constitutional language is hardly unambiguous and that the framers' intent on this matter is not crystal clear. After all, Congress could, if it chose, enact each line of an appropriations bill as a separate bill. The president would then, in effect, have an item veto. If that is possible, why is Congress barred from passing a general statute saying that appropriations bills should be considered separable for veto purposes? Moreover, even if the item veto is hard to square with the original constitutional structure, changing times have often persuaded the Supreme Court and the society that constitutional values must be applied flexibly to accommodate situations unfamiliar to the framers. Finally, the notion that Congress could not delegate the item veto power to the president is hard to understand in light of the broad delegations of legislative authority to administrative agencies. In sum, proponents argue that the item veto is a constitutionally appropriate response to the modern tendency towards massive appropriations bills and budget deficits.[10]

The item veto is not so clearly unconstitutional that it can be written off on the grounds that if it were ever enacted the courts would strike it down. Moreover, there is one decidedly legal way to enact the item veto—through constitutional amendment. While that is a long road, the near-success of the balanced budget amendment suggests that public dissatisfaction with the deficit could bring the item veto into reality one way or the other.

Under the circumstances, it is worth considering what the presence of the item veto might mean for government support for research in an area like serotonin. Analyzing that question requires an overview of how science spending policy is formulated in this country.

Although the federal government provides the bulk of the support for basic science research, there is no Department of Science and thus no single science budget. Basic research is funded primarily through the activities of the mission agencies, such as Defense, Energy, Health and Human Services. The National Science Foundation supports research in a variety of areas, but its overall contribution is smaller than that of the mission agencies.[11] Thus, today, with serotonin, research is spread throughout the federal government with grants and contracts being awarded by various agencies including the Public Health Service, the Department of Agriculture, the Department of the Army, and the National Institute of Mental Health.[12]

The absence of a Department of Science has generally been viewed as a healthy thing for American science. No one government official or agency can determine finally what is a valuable line of basic research. If research is cut off in one agency, it might survive in another. Given the uncertain progress of basic research, this sort of diffusion of effort can be an important safeguard.[13] Moreover, overall levels of science spending remain high compared to other countries. Basic research is, as a general proposition, popular with both liberals and conservatives. Even in difficult budgetary times, science spending does fairly well. In the Reagan years, for example, while research in civilian agencies often suffered, research in defense agencies often made up the difference.[14]

Given this background, the item veto does not threaten the basic relationship between the federal government and science. There is no reason to suspect that its presence would alter basic attitudes toward research. Moreover, even in those states where it is available, the item veto has generally been used to trim only 3 to 5% from appropriations bills.[15]

But there is reason to suspect that the item veto might pose a danger to certain specific types of scientific research. While basic research as a general concept is undeniably popular with most political figures, specific research projects sometimes suffer a different fate. Because research can seem esoteric and pointless to a lay observer, it is sometimes profitable for a politician to make fun of a particular project that might, in fact, be a valuable one. The item veto would give one very important politician—the president of the United States—the opportunity to score a few political points by vetoing a research plan that would strike the taxpayers, at least at first glance, as a waste of their money. In short, the "boondoggles" President Reagan identified as targets of the item veto might include some science.

Is serotonin research particularly subject to this sort of threat? It is impossible, of course, to predict with much certainty what might happen if an item veto came into existence. But there is some evidence to suggest that at least some types of research related to serotonin might suffer under an item veto regime. For while we have never had an item veto, we have had Sen. William Proxmire's "Golden Fleece" awards. Since 1975, Senator Proxmire has been giving these monthly awards for what he regards as "ridiculous, ironic and wasteful spending" by government agencies, and although the senator has now retired, he has vowed to continue to bestow the awards.[16] If, as seems likely, the senator's instincts in giving these awards are similar to some of the instincts a president might have in using an item veto, many scientists would do well to get nervous. Several of the "Golden Fleece" awards have been given to scientific researchers; indeed, even Proxmire admirers have described some of the awards as "Philistine."[17]

A particularly striking analogy to certain serotonin research is presented in the "Golden Fleece" Senator Proxmire gave in April 1975 to various federal agencies for supporting the research of Ronald Hutchinson.[18] Hutchinson, a research behavioral scientist, was studying primates to develop objective measures of aggression. He emphasized behavior patterns such as jaw clenching. In bestowing his award, Proxmire said:

> Dr. Hutchinson's studies should make the taxpayers as well as his monkeys grind their teeth. In fact, the good doctor has made a fortune from his monkeys and in the process made a monkey out of the American taxpayer.
>
> It is time for the Federal Government to get out of this "monkey business." In view of the transparent worthlessness of Hutchinson's study of jaw-grinding and biting by angry or hard-drinking monkeys, it is time we put a stop to the bite Hutchinson and the bureaucrats who fund him have been taking of the taxpayer.[19]

It does not take much imagination to picture a president characterizing serotonin studies on primates this way and then ending those studies with the stroke of the pen through the item veto.

None of this suggests that the item veto ought to be opposed by everyone and at all times. Science, like other areas of the budget, is hardly free of boondoggles. If a technique such as the item veto unavoidably harms a few valuable projects while lopping off many programs that deserve termination, it may be necessary to take the bitter with the sweet. But it is as yet unclear whether a federal item veto will in fact yield more benefit than harm. In undertaking that vital analysis, it is useful to keep in mind that genuinely valuable scientific studies might be unusually suspect to attack under an item veto regime.

NOTES

1. *See generally* S. Goldberg, *The Reluctant Embrace: Law and Science in America*, 75 GEO. L.J. 1341 (1987).

2. U.S. CONST. art. I, sec. 7.

3. At least that has been the assumption since the earliest days of the Republic. *See* C. Bellamy, *Item Veto: Shield Against Deficits or Weapon of Presidential Power?* 22 VAL. U. L. REV. 557, 571–574 (1988).

4. *Id.* at 558.

5. STAFF OF HOUSE COMM. ON RULES, 99TH CONG., 2D SESS., ITEM VETO: STATE EXPERIENCE AND ITS APPLICATION TO THE FEDERAL SITUATION 36–37 (Comm. Print 1986).

6. This statement, along with similar expressions of President Reagan's views, are cited in Bellamy, *supra* note 3, at 557 n. 3.

7. *See* P. Wolfson, *Is a Presidential Item Veto Constitutional?* 96 Yale L.J. 838 n. 2 (1987). Policy arguments on both sides are collected in *Symposium on the Line-Item Veto,* 1 NOTRE DAME J.L. ETHICS & PUB. POL'Y 157–283 (1985).

8. P. Holt, *Mr. Bush's Line-Item Veto,* THE CHRISTIAN SCIENCE MONITOR, September 5, 1991, at 18 col. 1.

9. *See generally* Wolfson, *supra* note 7; R. Spitzer, *The Item Veto Reconsidered,* 15 PRES. STUD. Q. 611 (1985).

10. *See generally* J. Best, *The Item Veto: Would the Founders Approve?* 14 PRES. STUD. Q. 183 (1984); G. Robinson, *Public Choice Speculations on the Item Veto,* 74 VA. L. REV. 403, 406 (1988).

11. Goldberg, *supra* note 1, at 1352–1354.

12. A search of computer data bases reveals scores of government-supported projects relating to serotonin. For example, the Dialog, Federal Research in Progress File 265 includes Neural Responses to Disproportionate Amino Acid Diets: Role of Monocemies (supported by the Department of Agriculture), the Dialog NTIS File 6 includes Aggression and the Biogenic Amine Neurohumors (supported by the Department of the Army and the National Institute of Mental Health); MEDLINE File 154 includes Factors Affecting Serotonin Uptake into Human Platelets (supported by the Public Health Service).

13. Goldberg, *supra* note 1, at 1354.

14. J. AREEN, P. KING, S. GOLDBERG, & A. CAPRON, LAW, SCIENCE AND MEDICINE, 1989 SUPPLEMENT 148 (1989).

15. *See e.g., Line-Item Veto: Hearings on S.J. Res. 26, S.J. Res. 168, and S. 1921 Before the Subcomm. on the Constitution of the Senate Comm. on the Judiciary,* 98th Cong., 2d Sess. 57 (Heritage Foundation Submission); 183 (Testimony of Governor Thompson) (1984).

16. J. Filas, *Personalities,* WASHINGTON POST, November 25, 1988, at D3, col. 4.

17. G. Will, *Proxmire's Record,* WASHINGTON POST, November 27, 1988, at D7, col. 2. *See also* S. Goldberg, *Controlling Basic Science: The Case of Nuclear Fusion,* 68 Geo. L.J. 683, 707 (1980).

18. *See* Hutchinson v. Proxmire, 443 U.S. 111 (1979). This litigation concerned Hutchinson's action for defamation against Senator Proxmire. Of course, no such action would be available if a president terminated a program through the use of an item veto.

19. *Id.* at 116.

16

Conclusions for Public Policy
Early Intervention, Special Education, and the Law

Roger D. Masters

Abstract: The scientific findings discussed in this volume will inevitably influence our legal system. Care is needed to avoid unintended consequences. The case is made that we should treat abnormally low serotonergic function as a claim for special education in order to minimize deviant behavior and enhance social responsibility. Such an approach would avoid many of the problems associated with introducing neurochemical evidence in the criminal law.

Extraordinary advances in research on the neurochemistry of behavior have created an urgent problem: how can such scientific discoveries best be integrated in our legal system? It is therefore fitting to conclude this volume by discussing the policy implications of research on the behavioral effects attributed to differences in serotonin function. Practical proposals are by no means easy to outline: in many respects, as is suggested by Rodgers (chapter 13), the law typically evolves solutions to new problems over time, and in so doing, the conscious intentions of lawyers, legislators, and scholars often do not determine the outcome (cf. Elliott, 1987).

At the same time, the way that scientific knowledge is related to our legal norms and public policies may make a significant difference. In paleontology, Stephen Jay Gould points out that small or accidental differences early in a process of change can have enormous, irreversible effects on subsequent evolution (Gould, 1989); similar effects have been noted in applying chaos theory to economics (Arthur, 1990). Given the disquieting nature of the discovery that much human behavior is influenced by neurotransmitters like serotonin, those with a sophisticated understanding of the law would do well to consider how best to apply the findings described in this volume.

The following conclusions are, of course, highly personal. Controversy is of the essence in issues of public policy. But it can hardly be controversial to hope that the introduction of scientific knowledge concerning the neurochemistry of behavior will not undermine the principles of our legal system. Do our basic notions of legal responsibility need to be changed? How can the discovery of the physical mechanisms shaping human behavior be integrated with the legal and moral traditions that were based on the notion of freedom and free will?

I. THE NEUROCHEMISTRY OF BEHAVIOR AND THE LAW

To address the legal and policy implications of research on the neurochemistry of behavior, it is important to begin from the nature of the legal system. Scientific findings will not, by themselves, suggest appropriate legal concepts and policy changes. As is implied by the divergence between Jeffery (chapter 11) and Shapiro (chapter 12), quite distinct legal approaches can be consistent with the proposition that neurotransmitters like serotonin play an important role in many human behaviors.

Nothing is more dangerous than the assumption that legal concepts should be viewed as descriptive categories that "ought" to reflect, directly and simply, scientific knowledge about the "objective" character of human behavior. Law is a symbolic system, regulating conflicts and tensions between norms of behavior and actual practices (or between conflicting norms). To ignore the difference between law and science can produce highly counterproductive results.

A simple example will illustrate the point. Lionel Tiger has pointed out that contemporary American legal and business practices take an apparently absurd approach to pregnancy. When an employed woman seeks a leave of absence for the purpose of childbirth, her condition is typically viewed as a "disability." Given the central role of reproduction in evolution, from a biological perspective one could consider childbirth as the principal "ability" of a woman. Why then do we consider a leave of absence for this purpose *disability* leave?

Reflection indicates a logic in the apparently illogical terminology. In our social system, there has often been extensive discrimination against women in the work force. Since males have occasion to seek disability leave but do not get pregnant, considering childbirth under this category may decrease the likelihood of discrimination against women leaving jobs temporarily to have babies.

As the example indicates, the meaning of a legal concept depends on the existing web of law in which it is embedded as well as on the social practices to which it applies. Law, as Elliott (1984) has put it, can be viewed as a system of artificial intelligence monitoring those areas of social behavior producing conflict or uncertainty. And as legal systems evolve in response to changing circumstances, the way of integrating scientific discoveries with the doctrines and practices of the law need not be governed by first appearances.

The unsuspected consequences of information about human nature are underscored by Rodgers's account of past practices by legislatures and courts. Politicians, legislators, and judges are subject to human passions; often, laws and legal decisions are framed in deceptive terms that further short-term goals. It would be most naive to assume that scientific findings concerning the role of neurotransmitters would be exempt from the general human tendency to use technical information in the light of political and social commitments of a quite different order.

There are, thus, both practical and theoretical reasons to prefer an element of indirectness in the relationship between science and legal theory. The behavioral effects attributable to serotonin or other neurotransmitters may not be most relevant to the legal doctrines concerning the causation of behavior. On the contrary, I would suggest that this is precisely the wrong area of legal doctrine to consider when introducing the information described in parts 2 and 3 of this book.

At first, evidence that human behavior is influenced by the baseline level or activity of a neurochemical such as serotonin seems to be "about" the problem of responsibility. If violent criminals or other undesirable or deviant behaviors seem to be associated with "abnormally" low levels of serotonin, it is tempting to conclude that the neurochemistry of behavior is related to questions of responsibility, particularly as they arise in criminal law. But as Shapiro (chapter 12) points out, there is good reason to hesitate before making this step. All legal systems presume that behavior is, in some sense, "caused." Were this not the case, there would be no reason to make laws, for laws are supposed to be one of the causes shaping future behavior. We recoil from holding individuals responsible when the actions in question were absolutely impossible to modify and totally independent of any decisions or choices of human agents, but we do so because it is assumed that the law is one of the causes that could change future behavior (either for the accused or for other members of society).

This is not to dismiss out of hand Jeffery's argument in chapter 11 that crime can be viewed in a sense as a medical condition. Sometimes one can hold individuals responsible for an illness: after being warned to avoid a high cholesterol diet, a man with a heart condition who ignores his physician's advice may be to some degree "at fault" when suffering a heart attack. Perhaps the medicalization of crime proposed by Jeffery could be quite consistent with traditional concepts of responsibility, even for criminals with abnormal neurotransmitter function, as Shapiro suggests.

In medicine, discovery of the causes of a disease becomes most relevant for prevention. As is often noted, many of the principal functions of law arise from its normative patterning of the behaviors that are *not* explicitly considered by courts or legislatures. In an important sense, as Wexler points out in chapter 14, legal concepts are a form of prevention, seeking to minimize the overt conflicts and confusion that can arise in complex human social interactions.

If so, we would do well to consider the complexity of information about serotonin and social behavior without assuming that it is primarily relevant to the question of human responsibility and crime. On the contrary, I will argue that the concepts of guilt, responsibility, and punishment are areas that should not be directly modified by the scientific findings discussed in the book. Instead, I suggest that we will need to look at the impact of serotonin on social behavior from the perspective of entitlements rather than punishments.

To be more specific, I propose that low levels of serotonergic functioning be considered in the context of rights to special education. In this view, biochemical imbalances should not be viewed as grounds for assuming diminished responsibility, for the very knowledge of the effects of neurochemistry—particularly in an age of sophisticated medical treatment—provides ample means for those who are affected to control the outcome (see Wilson, 1991).

It follows that the knowledge of an individual's neurochemistry could probably best be introduced into our system of law as a matter of entitlement to which the affected individual has a positive claim. Such information may, of course, also be relevant in the disposition of criminal cases, insofar as conditions of probation or incarceration have traditionally been modified in the light of knowledge about the individual's probable future behavior. But biochemical—and especially genetic—predispositions should *never* be accepted as exculpatory evidence, relieving individuals completely from guilt and responsibility for their actions.

To see the reasons for this proposal, it is necessary to summarize the scientific findings. In so doing, however, a simple fact should be remembered. Much more will be known about the neurochemistry of behavior at the moment that this book is being read than is known as I write this conclusion. Scientific knowledge is advancing at a rapid pace. Not only does this mean that we will have new information about the behavioral implications of genetics, individual experience, and neurochemistry, but also it means that we will know more about how to *alter* factors that seemed, even a few years ago, to be fated and beyond conscious human control.

If law is a system of artificial intelligence that evolves in response to changing circumstances, surely one should propose conceptual changes leaving room for future change. The suggestion that entitlements—and especially entitlements for special education—be the point of entry for neuroscience is made in this light. Entitlements are claims, made by those for whom they can be expected to be beneficial. No one is *required* to make such a claim. Starting in this area is thus more likely to contribute to progressive change as our society learns more about the physical causes of moral action.

Criminal law concerns collective passions: retribution, fear, and punishment. Introducing knowledge of the chemistry of behavior in an area of such strong passions may be most dangerous. If chemistry can cause illegal behavior, chemistry can also be used to produce social and political conformism. Given the changes in legal doctrine noted below, careful consideration needs to be given to the way we describe scientific findings about neurochemistry and behavior. Only in this way will we leave space for the law to evolve in ways that are consistent with our constitutional traditions.

II. SEROTONIN AND BEHAVIOR

The complexity of the human central nervous system is highly disconcerting. It can be predicted that many lawyers will have skipped Yuwiler, Brammer, and Yuwiler's survey of our knowledge about serotonin and its functional relationships (chapter 4). Such technical information often seems irrelevant to the basic legal issues. Nothing could be more dangerous as neurochemistry continues to expand our understanding of the causes and control of behavior.

The very complexity of the neurochemical systems outlined by Yuwiler, Brammer, and Yuwiler helps to explain the seemingly paradoxical character of a single neurotransmitter such as serotonin. On the one hand, specific behavioral deficits or physical illnesses can be directly attributed to disturbances in serotonergic functioning; on the other, serotonin is never entirely responsible for a specific outcome, and low levels of serotonin can be related to a variety of different behavioral conditions.

The basic rule of thumb can be restated as follows: a given behavior can have several different causes; a given cause can give rise to several different behaviors. This rule of thumb is more complicated than simplistic cause-effect or stimulus-response "laws of nature." But complexity is not random or unpredictable confusion. Patterns exist. Simple determinism does not, however, describe the facts.

In the future, we can expect further discoveries of individual differences in genetic predispositions that are associated with behavior (Plomin, 1990). In recent months, much has been said of a genetic predisposition to alcoholism. As the research of Linnoila and his colleagues indicates, such predispositions often seem to be related to neurochemical imbalances. Indeed, as Linnoila points out, some alcoholism may be viewed as a crude attempt at self-medication insofar as the short-term effects of drinking may be to release serotonin in individuals with low baseline levels of this neurotransmitter.

The genetics underlying behavioral neurochemistry thus concerns vulnerabilities, not necessities. As Ginsburg and his colleagues have shown in other mammals (chapter 8), social experience and individual physical health or development may trigger genetic dispositions that are otherwise never realized. Such individual differences may help to explain the apparently paradoxical results of serotonin found in Madsen's experimental findings with humans (chapter 10). And the more one knows about the effects of neurochemistry on behavior, the more these effects can be controlled by changing behavior as well as by medication.

To understand how these discoveries might influence law and public policy, it is imperative that the main effects of serotonin be described accurately. At first, it might seem that individuals with low levels of serotonin are likely to be more aggressive and violent. The analysis of the role of serotonin in suicide by Stein and Stanley (chapter 5) indicates that this is misleading: low levels of serotonergic functioning seem to produce deficits in impulse control rather than aggressiveness or violence.

The remarkable findings of Linnoila and his colleagues, using measures of serotonin and glucose uptake to predict recidivism among those guilty of arson and homicide (chapter 6), need to be considered in this context. Not only is the deficit attributable to neurochemistry associated with impulse control rather than with violence per se, but also the actions characterized as criminal are typically committed by individuals who are under the influence of alcohol or drugs at the time of the crime. Isolating information concerning the probable role of serotonergic imbalance from the overall understanding of the neurochemistry of behavior is all the more misleading, for part of that neurochemistry is attributable to patterns of alcoholic consumption.

In assessing the relevance of such discoveries for our notion of legal responsibility, moreover, it is well to keep in mind other behavioral syndromes associated with similar neurochemical imbalances. While serotonin and glucose uptake seem related to impulsive violence, the interaction of serotonin and melatonin is implicated in seasonal affective disorder (SAD). As Wurtman and Wurtman show (chapter 7), the discovery that levels of illumination are the environmental trigger for seasonal depression makes possible inexpensive and reliable therapies and means of modifying neurochemical function that need not entail the use of medications and drugs. Much the same can be said of the role of serotonin in the symptoms of premenstrual syndrome (PMS).

The discovery of physiological and even genetic causes of seasonal affective disorder or PMS thus increases the ability of those affected to control outcomes. There is good reason to treat the forms of impulsive violence analyzed by Linnoila and his colleagues in the same way. Many of the prisoners convicted of impulsive homicide or arson are models of behavior during incarceration: being compelled to abstain from alcohol by the conditions of prison life, they seem in perfect self-control. Predictions of recidivism based on conventional measures of behavior while incarcerated are thus unreliable, whereas an understanding of the role of alcohol can greatly increase the expectations of responsibility.

The perspective outlined here is in equally sharp contrast to the traditions of both behaviorist psychology and genetic determinism. According to traditional behaviorism, a stimulus or environmental condition is presumed to have a constant effect on behavioral response, subject only to differences in the experience or conditioning of the individual. According to genetic determinism, a gene can code "for" a trait or behavior in ways that are generally independent of life experience. Modern behavior genetics shows that the same experiences often have different effects on individuals, depending on both genetic predispositions and early experiences. On the other hand, this means that the expression of genetic tendencies or predispositions depends on personal life experiences and on concrete circumstances under individual control.

Several crucial implications follow. First, for the conditions described in this book, one cannot predict adult behavior from an individual's genes at birth. Not only can prenatal and infant development condition the expression of the genes involved, but knowledge of these gene-environment interactions can itself lead to modifications of behavior (Wilson, 1991). *The more we know about the genetics of behavior, the greater the responsibility of those affected by genetic predispositions to deviant behavior.*

Second, for many of the conditions described in this book, early childhood is critical. Serious disturbances of mother-infant (and perhaps father-infant) bonding may be dangerous for all children. In the case of those with genetic predispositions to low serotonergic functioning, hypoglycemia, and alcoholism, such developmental "insults" can have the effect of triggering a latent vulnerability. While the exact neurochemical mechanisms involved are not known, the principle is adequate for questions of public policy and legal doctrine: *insofar as the tendency to suicide or impulsive violence is associated with serotonergic dysfunction, this relationship is in part sociogenic.* That is, we can consider that society has as much of an obligation to help those afflicted as it does to help children suffering from an epidemic of measles or polio.

Serotonergic dysfunction is associated with many behavioral traits, most of which are not in any way criminal. Even for the behaviors contributing to criminal violence, serotonergic dysfunction typically will have behavioral effects long before a child has left school. If so, the medicalization of crime described by Jeffery would seem to be best understood as the basis of a claim to special education and other forms of entitlement that would address the affected individuals early enough in development to modify lifetime careers.

The patterns associated with impulsive violence due to serotonergic dysfunction are often manifested as disturbances in the behaviors associated with normal education during childhood. Attentional disorders and poor learning are often found in children with deficient impulse control. Poor learning often leads to marginal social status and failure. In turn, marginal social status and failure can contribute to impulsive violence and crime.

Instead of assuming that neurochemistry is evidence of diminished responsibility, one might well adopt the opposite view, according to which neurochemical imbalance in childhood would provide an entitlement to special educational opportunities, much as poverty has provided an entitlement to such benefits as subsidized food in the WIC program. In this view, full moral responsibility should be attributed to all adults: if neurochemistry is a contributing cause to the criminal behavior of some, abusive parents or poverty contribute to the criminal behavior of others. All behavior is caused. The law should not be based on indiscriminate pity.

III. SEROTONERGIC FUNCTION AS THE BASIS OF ENTITLEMENT

In suggesting that serotonergic dysfunction be viewed as an entitlement, it is important to work out the implications of using physiological or medical information about the neurochemistry of behavior as the basis of a legal claim or "right." Intuitively, many lawyers and criminologists have considered the effects of serotonin in terms of responsibility (and especially criminal responsibility). Before considering the potential disadvantages of this apparently "obvious" way of integrating neurochemistry of behavior into the law, we need a clearer picture of how the proposed approach might work.

In a number of its manifestations, low levels of serotonergic functioning seem to be associated with deficits in impulse control. What at first was classified as an increase of aggressiveness has, on closer inspection, been found to be an effect due to impulsiveness (Stein and Stanley, chapter 5). This helps to explain why similar biochemical characteristics can be associated with suicide, arson, and homicide (and notably homicide in which the name of the victim is not even known to the aggressor). Insofar as depression is interpreted as an inability to inhibit thinking about negative outcomes, the common element in many forms of low serotonergic function might be hypothesized as a loss of the ability to inhibit or control neuronal action.

Impulsiveness is not, in itself, criminal—or even deviant. The low levels of serotonin associated with some forms of criminal violence (chapter 6) are also implicated in conditions (PMS, suicidal behavior) that are generally viewed as medical or psychiatric. More important, for those young males with the combination of hypoglycemia and low serotonergic levels, many of the traits associated with these biochemical conditions have broader effects during childhood that may help explain the incidence of criminal behavior in later years among those affected. The proposal that neurochemical imbalance be the basis of claims for specialized educational or medical treatment in childhood will probably not be controversial in cases where the effects are already viewed in this manner.

Let us focus, then, on the population at risk for the form of criminal behavior and recidivism studied by Linnoila and his colleagues in chapter 6. Such individuals are often males with a history of familial alcoholism, poor familial bonding, difficulties in impulse control, educational failure, and social marginality. Each of these factors can contribute independently to the others. How, then, can the vicious circle be broken? To take an extreme case, where can or should society act to protect itself against the consequences of psychological neglect or abuse of a genetically vulnerable infant by alcoholic parents, producing educational deficits, deviance, and ultimately criminal actions?

In our legal system, the normal point of generalized social responsibility for the development of citizens occurs at the entry to the public school system. There are, of course, many conditions in which we justify legal intervention prior to this point: physical child abuse, for example, is generally considered grounds for state action against parents. But it is hardly conceivable and surely undesirable to establish effective mother-infant bonding as a legal requirement subject to judicially enforced punishments; the consumption of alcohol—or even diagnosis as an alcoholic—is not itself criminal. The protection of privacy and the dangers of an uncontrolled expansion of state intervention in the lives of citizens surely militate against a punitive approach to the earliest stages of the pathway leading to the syndrome of hypoglycemia, low serotonergic functioning, deficits in impulse control, and criminal behavior.

Within the school context, in contrast, the behavior of the child is an appropriate and indeed a necessary focus of legal concern. Not only does each individual's behavior have effects on others, but society has—by the very fact of establishing public education as a mandate—accepted responsibility over a portion of the development and maturation of its citizens. Bearing this in mind, let us consider how the neurochemical conditions that, for some adults, might be associated with criminal behaviors are manifested in school-age children.

Deficits in impulse control are well known to elementary school teachers. Although "hyperactive" children often suffer from educational deficits, perhaps it is prudent to set aside a category that might be extended to any child engaging in bursts of physical activity. But many forms of uncontrolled impulsiveness fall within the established categories of learning disabilities, considered as the basis for entitlements to special education because they are associated with impairments in normal cognitive development.

Attention deficit disorder (ADD), for example, is now generally recognized as a condition or set of conditions in which children are likely to exhibit organically based deficits in the ability to attend to the learning tasks of the classroom. It is neither necessary nor possible to explore the extensive literature on this category of learning deficits: virtually nothing is known, for example, about their relationship to the genetic or neurochemical substrates of behavior discussed in this volume. Even if attentional deficits can be attributed to a wide variety of causes, however, it is plausible to consider the impulsiveness associated with low serotonergic function and hypoglycemia as a potential element in poor attention, troublesome social behavior, and ineffective learning in the school context.

It is hardly necessary to stress the particular importance of cognitive skills for those who may be at risk for social marginality. From a biological perspective, there is a group of children who are genetically vulnerable to behaviors that are either self-destructive or socially destructive, but the expression of this tendency depends in part on early childhood experiences. Even if some such individuals may develop relatively irreversible behavior patterns by adulthood (and of course many would still contest this possibility), the current understanding of organic development indicates that early intervention is both physiologically feasible and socially necessary if these tendencies are to be prevented.

The claim to special attention in early education is, at present, possible in cases of cognitive deficits. Dyslexics, for example, are usually entitled to special educational programs if their performance on specific language tests is one and one-half standard deviations below general measures of intelligence and mental ability. In many cases, moreover, medical diagnosis of learning disability is based in part on neurological dysfunctions now known to be associated with language deficits. Similarly, of course, the hearing- or sight-impaired have claims to appropriate services. In these cases, American law has accepted cognitive and neurological measurements as the basis of entitlements to specialized educational services.

By treating low serotonergic functioning as a potential basis for claims to special education, therefore, one could insert new information on the biochemistry of behavior in an established area of the law that does not entail questions of changing our society's basic concept of legal "responsibility." In itself, this may be welcome: whereas passions run high in questions of the guilt and punishment for violent crime, educational policy may be a more reasonable domain in which to introduce the extraordinarily complex scientific findings concerning the neurochemistry of behavior.

Many other advantages flow from this strategy. By viewing individual neurochemical balances as an entitlement to special benefits, the introduction of evidence concerning serotonin and behavior would arise at the option of the individual involved. As a result, claims and counterclaims concerning the actual effects of any specific neurochemical condition would depend on the willingness of affected individuals and their families to make claims. Because of the highly variable character of special educational planning (in general, each student concerned must have an "individual educational plan" [IEP] agreeable to both parents and school), standards could evolve as more is known about the precise relationships between complex neurochemical patterns and specific behavioral or personality traits.

In contrast, considering neurochemistry primarily in terms of criminal law could lead to the premature establishment of precedent, thereby expanding the practice of imposing drug treatment by the judicial system. The potentially coercive effects of knowledge about an individual's biochemistry could be minimized by focusing on civil entitlement rather than criminal law. The emphasis would be on the prevention of social disability and deviance rather than on the use of medical knowledge to determine guilt, innocence, and punishment. If special education programs were successful, children with abnormal biochemical profiles would improve learning and be more likely to behave in socially acceptable ways.

More important, the evidence from Raleigh and McGuire's study (chapter 9) suggests that replacing repeated school failure with the positive social interaction might itself alter neurochemistry. It has often been noted that children with such learning disabilities as attention deficit disorder are in a vicious cycle of failure and negative reinforcement. Special education often makes it possible to break this cycle, producing positive social reinforcement for the first time. Insofar as effective integration in a meaningful social group might contribute to higher serotonergic levels, the proposed approach might be physically as well as educationally beneficial to affected children.

To be sure, there are risks entailed in introducing neurochemical or neurophysiological information to the educational process. Few classroom teachers are equipped to understand the latest scientific findings on the structure and function of the central nervous system. Because the prevailing educational philosophies emphasize the equality of all children, the discovery that behavioral and educational performance has a biological substrate could lead to abuses in the school setting.

Many would fear the effects of "labeling" or stigmatizing the affected children. If neurochemistry is viewed as an independent "cause" of behavioral and educational difficulties, perhaps it would be assumed the medication would resolve deficits that actually require careful social and pedagogical intervention. Special education is costly, and our school system is already underfunded. Widely different remedial programs may be needed for dyslexia, attention deficit disorder, and emotional disturbance. Teachers generally lack the expertise to diagnose different forms of learning disabilities.

Without discounting these concerns, it can hardly be denied that information about the neurophysiology and neurochemistry of learning will continue to develop rapidly in coming years. Researchers have claimed to find genes associated with traits as diverse as schizophrenia, alcoholism, and dyslexia. Even if such findings are not conclusive at this writing, it is only prudent to assume that more will be known about the genetics of behavior and learning in coming years. Evidence of prenatal influences on development that modify learning and behavior may be equally important in the discovery of individual differences in learning potential.

In the three decades after World War II, American education was dominated by behaviorism and the assumption that environmental factors were totally responsible for differences in educational performance. Increasingly, our school systems will be confronted with scientific evidence that individuals differ. Only slowly have educators begun to realize the importance of neurologically based learning disabilities and the need for special education programs designed to meet neurologically based deficits that take different forms from one child to another.

In recent years, special education has begun to reduce the effects of dyslexia on educational attainment and career success by providing "coping strategies" and positive social experiences for children who have known only failure. Far from encouraging an attitude of genetic determinism and fatalism, recent scientific research points to the interaction between biological and social factors in the development of social competence, learning, and the capacity to fulfill normal adult roles. If so, our schools will in any event be moving toward the acceptance of what Howard Gardner has called the theory of "multiple intelligences": in place of the presumption of a single pattern of learning appropriate to all children, contemporary biology suggests the heterogeneity of learning abilities. Since children need distinct modes of experience, creating entitlements for those with a biological predisposition to attention deficit disorder or other neurochemical imbalances associated with impulsive and violent behavior would embed information concerning the neurochemical determinants of behavior in a positive context more likely to be controlled by citizens and to produce a salutary evolution of legal doctrines.

IV. THE DANGERS OF "DIMINISHED RESPONSIBILITY"

If it is plausible to consider low serotonergic function in the context of educational disability, need this exclude the view that those exhibiting impulsive violent behavior related to neurochemical causes have a "diminished responsibility"? Isn't it only fair to consider that individuals whose criminal behavior can be predicted on the basis of chemical tests are not fully in "control" of their actions and hence should not be considered guilty if they have committed criminal acts? To answer this question, it is important to bear in mind three distinct factors: the logic of the concept of responsibility, the complexity of the scientific findings, and the broader legal and social environment within which these events occur.

A. The Concept of Responsibility

The argument sketched above suggests that an extension of the logic of the "insanity defense" to allow neurochemical claims for "diminished responsibility" is unwarranted and unwise. Indeed, the logic of the putative defense is backward: at the individual level, the more that is known about the biochemistry of behavior, the more one can legitimately hold responsible those individuals affected by genetic or neurological predispositions toward deviant behavior. If hypoglycemics with low levels of serotonin are more likely to commit crimes when under the influence of alcohol or drugs, knowledge of these facts should *increase* (not decrease) the culpability of individuals who have been diagnosed to be at risk. The reasons are self-evident when one considers purely medical conditions without a legal dimension: in seasonal affective disorder, for example, discovery of the neurochemistry of behavior has provided effective responses such as light therapy (Wurtman and Wurtman, chapter 7).

Neurochemistry is not fixed and impervious to behavior, diet, or medication. Our norms of moral and legal responsibility need not be changed in the light of the research discussed in this volume (Shapiro, chapter 12). Whether or not one agrees with the argument that the insanity defense itself should be abandoned, it seems unwise to use neurochemistry as a legal defense against criminal guilt.

It will be objected that it is unfair to hold someone responsible for a biologically based tendency to criminal behavior when under the influence of alcohol. Because many people can have one or two drinks without encountering serious problems of impulse control, doesn't the concept of equality before the law contradict such a policy? John Stuart Mill's *On Liberty*, a work now generally viewed as the classic defense of the individual against arbitrary control by society, answers this objection directly.

Because the principle of legal responsibility must take into account individual circumstances, those who know they are susceptible to impulsive behavior while under the influence of alcohol or drugs can be held accountable for consuming substances that, for others, are legal and permissible. In *On Liberty*, Mill refers explicitly to this issue with regard to alcoholic beverages. Since his argument applies well to the criminals described in chapter 6, it is worth citing in detail.

Mill strongly condemns the attempt to prohibit the sale or consumption of intoxicating beverages, which he uses as an example of "illegitimate interference with the rightful liberty of the individual" (Mill, 1956 [1859]:110). Long before the failure of America's national experiment with prohibition, he stressed the dangers of imposing decisions on the use of alcohol by law. Insofar as the mere fact of being intoxicated is a "self-regarding" action that harms only the drunkard, society is not properly justified in punishing the sale or use of alcohol (Mill, 1956:108–9, 115–19).

How, then, can society protect itself against those whose abuse of alcohol is related to repeated criminal acts? For Mill, although the mere fact of being drunk cannot usually be the ground of criminal prosecution, those who have once demonstrated criminal impulsiveness while drunk can—and should—be held responsible for drinking *even before any new criminal act has been committed*.

> Drunkenness, for example, in ordinary cases is not a fit subject for legislative interference, but I should deem it perfectly legitimate that a person who had once been convicted of any act of violence to others under the influence of drink should be placed under a special legal restriction, personal to himself; that if he were afterwards found drunk, he should be liable to a penalty, and that if, when in that state, he committed another offense, the punishment to which he would be liable for that other offense should be increased in severity. The making himself drunk, in a person whom drunkenness excites to do harm to others, is a crime against others. (Mill, 1956:119)

Mill's principle could be described as *enhanced* responsibility. The more one knows about the causes of one's behavior, the greater the reason to expect this knowledge to be used to avoid criminal acts. This view is consistent with legal precedents, such as the case of *People v. Tocco* cited by Wexler in chapter 14. As James Q. Wilson (1991) has argued, failure to adopt such a principle is tantamount to abandoning civilized standards of behavior, which are based on the premise that humans can and should develop self-control when confronted with potentially harmful impulses and desires.

B. The Complexity of Neurochemistry

A second reason for rejecting the use of low serotonergic function as evidence of diminished responsibility arises from the complexity of the scientific findings. As

Yuwiler, Brammer, and Yuwiler (chapter 4) point out, the chemistry involved is exceptionally complex. Even if the scientific theories are compelling, there are practical difficulties in establishing unambiguous measures of neurotransmitter function that are unaffected by temporary physiological states and the coactivity of other chemical compounds active in the central nervous system.

Serotonin levels alone are, quite obviously, not responsible for crime. On the one hand, we have seen that low serotonergic function is associated with a number of conditions, including depression, seasonal affective disorder, premenstrual syndrome, and suicide; even if serotonin is one factor in explaining a specific category of criminal behavior, the precise effects depend on a combination of chemical systems. On the other hand, not everyone with the combination of organic predisposition to hypoglycemia and low serotonergic function engages in impulsive crime. Not only does individual experience also play a causal role, but those affected seem more capable of controlling their behavior if they abstain from alcohol and drugs.

C. The Legal and Social Context

Even if one dismissed the foregoing considerations, recent changes in the law suggest that viewing low serotonergic functioning in terms of criminal liability and guilt could have effects diametrically opposed to those one might expect. There has been a sharp change in legal precedents concerning the relationship between psychiatric treatment and punishment for crime over the last 15 years. Because medical decisions have become increasingly legitimate in the behavioral management and rehabilitation of convicted criminals, a neurochemical-based defense of "diminished responsibility" could well be an invitation to unlimited extensions of chemical control of behavior, not only in our prisons but also in society at large.

As recently as 1979, the Supreme Court overturned a Nebraska law that permitted the transfer of a prisoner to a mental institution if a physician or psychologist found the individual "suffers from a mental disease or defect" that "cannot be given proper treatment in prison" (*Vitek et al. v. Jones*, 445 U.S. 480–500). In that case, it was held that transfer to a mental hospital "without adequate notice and opportunity for a hearing" deprived a prisoner "of liberty without due process of law": "although the State's interest in segregating and treating mentally ill patients is strong, the prisoner's interest in not being arbitrarily classified as mentally ill and subjected to unwelcome treatment is also powerful, and the risk of error in making the determinations required . . . is substantial enough to warrant appropriate procedural safeguards against error" (*ibid.*, 480–81).

Since this decision, not only has crime and the treatment of criminals become a more visible political issue in the United States, but also the psychoactive drugs available for treating prisoners and mental patients have become more varied and more effective. As a result, the line of precedents since *Vitek et al. v. Jones* has moved progressively to a more permissive stance toward the forcible administration of drug therapy to convicted criminals.

In *Mills et al. v. Rogers et al.* (457 U.S. 291–306), the court addressed the question of forcible administration of antipsychotic drugs to *noncriminal* mental patients and remanded the case for determination of the exact balance between "the constitution-ally protected liberty and privacy interests" and "the identification of the conditions under which competing state interests might outweigh" the patient's rights. As applied to those accused or convicted of crime, courts often came down on the side of the prisoner's constitutional rights, as in *Bee v. Greaves* (744 F. 2d 1387 [10th Circuit, 1984]; 469 U.S. 1214 [1985]). But in the last few years, the balance has shifted.

In *United States v. Charters* (863 F. 2d 302 [4th Circuit, 1988]), "the court held that a person accused of a serious federal crime and committed, as incompetent to be tried, to the federal prison at Butner, North Carolina, has no federally protected right to resist the forcible administration of antipsychotic medication." More recently, the Supreme Court confirmed a similar ruling in *Harper v. Washington* (1990).

As this trend of court cases indicates, the adoption of a concept of "diminished responsibility" by reason of individual neurochemistry is likely to change the pretrial treatment of accused criminals as well as the disposition of criminal cases. Well-intentioned defense lawyers may feel they are benefiting clients by reducing the formal admission of guilt. In fact, the results could well be both the legitimation of forcible medication before trial and the presumption of guilt in the judicial process.

V. CONCLUSIONS

In introducing this volume, I noted how knowledge can be dangerous, particularly in a constitutional government that depends on the educated judgments of its citizens, its judges and lawyers, and its legislators. The legal and scientific implications of research on the neurotransmitter serotonin, which have been the subject of the intervening chapters, are important not only in themselves but also as an illustration of a broader problem confronting our society. How can we live with more extensive knowledge concerning the biological causes of human behavior than has ever before been accessible to human beings?

Rapid advances in the study of the neurochemistry of behavior are extremely unsettling for the comfortable orthodoxies that have dominated our political rhetoric as well as our legal codes. Given the extensive changes in our understanding of behavior that are occurring, it does not seem to be an exaggeration to speak of "the neurotransmitter revolution." In place of the "blank slate" on which experience and environment can write any message of personality, attitudes, and behavior, contemporary biology is discovering a complex interaction of genetics, physiological development, and personal experience in which individual thought and choice is conditioned by a combination of social customs, individual learning, and individual differences in neurochemistry or neuroanatomy.

We have few models for an approach to these radically new scientific discoveries that can inform our legal process and cultural understanding while opening the consequences to informed debate. Our newspapers tend to focus on the dramatic "news," while science reporting is, given the awesome developments in our universi-ties and laboratories, woefully limited and incomplete. Worse yet, our leaders tend to emphasize the same short-term issues that dominate the headlines: legislators and judges lack the time and the expertise to deal with rapidly accumulating discoveries in the life sciences.

This book has sought to contribute to a more informed dialogue among the scientific community, the citizen, and the legal system. We have surveyed a variety of findings concerning serotonin as one of the major neurotransmitter systems in the human brain (parts 2 and 3), outlined the nature of the legal and moral issues (part 4), and illustrated the way the law as an evolving body of concepts, rules, and practices might be influenced by such information about human nature (part 5).

Similar considerations can be raised about a host of research areas in contemporary biology. Our legal scholars and law schools will need to open such subjects to a dialogue in which neuroscientists and ethologists communicate with criminologists, political scientists, and philosophers (to cite only some of the disciplines that need to come into contact if our civilization is to control the knowledge explosion instead of being controlled by it). This book is thus intended as a beginning, not an ending; as an illustration of a method and a process, not a definitive conclusion.

For example, I have outlined a way of integrating information about an individual's neurotransmitter levels into our legal system. Whereas the first temptations have been to introduce evidence of low serotonergic functioning as the basis of a claim of diminished responsibility in criminal law, I have proposed an alternative strategy: insofar as neurotransmitter activity can be convincingly associated with behavioral dysfunction, I suggest that such evidence be viewed as a matter of civil law, not criminal law, to be used as the basis of claims for entitlement to special education rather than as exculpatory evidence in criminal cases. If abnormally low serotonin level or activity enters into criminal procedure at all, it would best be limited to factual conditions associated with probation and sentencing rather than admitted as relevant to the legal and moral responsibility of the actor.

These proposals have not been advanced as the final word on the matter—quite the contrary, for this volume includes ample evidence of both the awesome complexity of the scientific findings involved and the unanticipated effects of innovation on the evolution of the law. Dialogue is necessary if a constitutional regime is to absorb such evidence without abandoning its concepts of legal rights and due process. Informed citizens, scientists, and legislators need to consider new ways of understanding human behavior with an open mind. Whether or not the reader agrees with the arguments set forth by the contributors, if we have contributed toward that end, our work will have been successful.

REFERENCES

Arthur, W. Brian. 1990. Positive Feedbacks in the Economy. *Scientific American* 262:92–99.

Bee v. Greaves, 744 F. 2d 1387 [10th Circuit, 1984]; 469 US 1214 [1985].

Elliott, E. Donald. 1984. Holmes and Evolution: Legal Process as Artificial Intelligence. *Journal of Legal Studies* 13: 113–45.

Elliott, E. Donald. 1987. What's Wrong with Our Constitutional Law? Paper presented to the "After The Bicentennial" Conference, Georgetown University Law Center, Washington, D. C., November 14–15, 1987.

Gould, Stephen Jay. 1989. *Wonderful Life*. New York: W. W. Norton.

Mill, John Stuart. 1956 [1859]. *On Liberty*. New York: Macmillan/Library of Liberal Arts.

Mills et al. v. Rogers et al., 457 US 291–306.

Plomin, Robert. 1990. The Role of Inheritance in Behavior. *Science* 248:183–88.

United States v. Charters, 863 F. 2d 302 [4th Circuit, 1988].

Vitek et al. v. Jones, 445 US 480–500.

Wilson, James Q. 1991. *On Character*. Washington, D. C.: AEI Press.

Notes on Contributors
Index

Notes on Contributors

Gary L. Brammer received his Ph.D. under Dexter French in the Department of Biochemistry and Biophysics, Iowa State University, before taking a postdoctoral fellowship in neurosciences at the University of California at Los Angeles. He is currently Associate Chief of Neurobiochemical Research at the Veterans Administration Medical Center, West Los Angeles, and Associate Research Biochemist, Department of Psychiatry and Biobehavioral Sciences, at UCLA.

Benson E. Ginsburg is Professor of Biobehavioral Sciences and Psychology at the University of Connecticut, Storrs, and Professor of Psychiatry at the University of Connecticut Health Center, Farmington. He received his Ph.D. from the University of Chicago, where he later became William Rainey Harper Professor of Biology. He was twice a fellow at the Center for Advanced Study in the Behavioral Sciences and has been a visiting scholar at the University of Cambridge and the Hebrew University of Jerusalem, where he was the Scheinfeld Professor of Human Genetics in the Social Sciences. His many publications are in the field of behavior genetics. One of his major research interests has been with studies of aggressive behaviors.

Steven Goldberg, J.D., is a professor of law at the Georgetown University of Law Center in Washington, D.C. A graduate of Harvard College and Yale Law School and a former law clerk to Supreme Court Justice William J. Brennan, Jr., Goldberg writes primarily in the field of law and science. He is coauthor of the text *Law, Science and Medicine*.

Margaret Gruter holds a Dr. jur. from the University of Heidelberg. She received an advanced degree (J.S.M.) from Stanford University Law School in 1973 for her work on the interaction between law and biologically based human behavior. She has taught at Stanford and Heidelburg universities and has lectured at the University of

Zurich. Her books and articles have been published in English as well as in German. In the early 1980s, she founded the Gruter Institute for Law and Behavioral Research, which examines relationships between law and biology, with an emphasis on how recent findings in biology and anthropology inform the history, practice, understanding, and utility of law. She continues her research together with research fellows of the institute. The Gruter Institute is the only one of its type in the world, and it continues to remain at the intellectual forefront of biology-law explorations.

C. Ray Jeffery is Professor of Criminology and Criminal Justice at Florida State University in Tallahassee. He is past president of the American Society of Criminology and recipient of the Sutherland Award from that society. He is founding editor of *Criminology*. He has also been a Fulbright-Hays Research Fellow in the Netherlands and was George J. Beto Professor at Sam Houston University.

Markku Linnoila is a native of Helsinki, Finland. After receiving an M.D. at the University of Helsinki (1972), he studied biochemistry and pharmaceutical chemistry, receiving a doctorate in pharmacology in 1974. Thereafter he was a research fellow and professor at the Duke University Medical Center, where he became Head of the Clinical Psychopharmacology Section in 1979. After serving as staff psychiatrist at the National Institute of Health, he became Chief of the Laboratory of Clinical Studies and Clinical Director of the Division of Intramural Clinical and Biological Research at the National Institute on Alcohol Abuse and Alcoholism (NIAAA). Since 1991, Dr. Linnoila has been scientific director of the NIAAA. He is author or coauthor of over 400 scientific papers and coeditor of *Risk Factors in Youth Suicide* (New York, 1991).

Michael T. McGuire, M.D. (University of Rochester), is Professor of Psychiatry and Biobehavioral Sciences at the University of California at Los Angeles. He is a member of the Brain Research Institute and is coeditor-in-chief of *Ethology and Sociobiology*. His primary research interests are behavior-physiology interactions among nonhuman primates and evolutionary theory as it applies to mental illnesses. Among his many publications are recent articles in *Developmental Psychobiology, Animal Behavior, American Behavioral Scientist,* and the *University of Southern California Law Review.*

Douglas Madsen, who received his Ph.D. from the University of California at Los Angeles in 1973, is Professor of Political Science at the University of Iowa. His research focuses primarily on social dominance and leadership. Recent publications include *The Charismatic Bond: Political Behavior in Time of Crisis* (Harvard University Press, 1991), written with Peter G. Snow, and various journal articles. Currently, he is involved in replicating and expanding the serotonin findings reported in chapter 10.

Roger D. Masters is Nelson A. Rockefeller Professor of Government at Dartmouth College and editor of the *Gruter Institute Reader in Biology, Human Social Behavior, and Law* (Primis-McGraw Hill, 1992). His latest books include *The Nature of Politics*

(Yale, 1989), the first three volumes of the *Collected Writings of Rousseau* (coedited with Christopher Kelly, University Press of New England, 1991–1993), and *The Sense of Justice* (coedited with Margaret Gruter, Sage, 1992). He is also coeditor, with Glendon Schubert, of *Primate Politics* (Southern Illinois University Press, 1991).

Michael J. Raleigh is an associate professor of psychiatry and anthropology at the University of California at Los Angeles. For more than a decade he has investigated the biological causes and consequences of social behavior. His research has focused on the bidirectional relationships between behavior and monoamine function. This work examines the contributions of central nervous system monoaminergic systems, including serotonin, norepinephrine, and dopamine, to the regulation of social and individual behavior in nonhuman primates. It also evaluates the constraints behavioral and environmental factors place on neural function. Dr. Raleigh is editor-in-chief of the *American Journal of Primatology*.

William H. Rodgers, Jr., J.D., holds the Bloedel Chair of Law at the University of Washington. He has been a member of the Board on Environmental Studies and Toxicology of the National Academy of Sciences and has lectured at a number of law schools. His book, *Environmental Law: Air and Water*, has been published in Japanese; he is also the author of a treatise on environmental law, a work in four volumes.

Michael H. Shapiro is the Dorothy W. Nelson Professor of Law at the University of Southern California Law Center. He teaches bioethics and law, constitutional law, and health care law, has taught criminal law and prisoners' rights, and writes on matters touching these fields. He received a B.A. and M.A. in philosophy from the University of California at Los Angeles and a J.D. from the University of Chicago and is a member of the California Bar.

Michael Stanley received his Ph.D. at The Mount Sinai School of Medicine in New York. Before his death in 1993, he was a professor of psychiatry and pharmacology at the College of Physicians and Surgeons of Columbia University. The focus of his research and scholarship was the biochemical basis of suicide and related forms of self-injury. Additionally, he wrote on issues of informed consent and competency in psychiatric patients.

Dan J. Stein studied medicine at the University of Cape Town. He completed his psychiatric residency at Columbia-Presbyterian Hospital and the New York State Psychiatric Institute, where he is currently a postdoctoral research fellow. He is a recipient of a NARSAD (National Alliance for Research on Schizophrenia and Depression) Young Investigator Award.

David B. Wexler is John D. Lyons Professor of Law and Professor of Psychology at the University of Arizona. His books include *Essays in Therapeutic Jurisprudence* (with Bruce Winick, Carolina Academic Press, 1991), *Therapeutic Jurisprudence: The Law as a Therapeutic Agent* (Carolina Academic Press, 1990), and *Mental Health*

Law: Major Issues (Plenum Press, 1981). He received the American Psychiatric Association's Manfred S. Guttmacher Forensic Psychiatry Award, chaired the American Bar Association Commission on the Mentally Disabled, was a member of the National Commission on the Insanity Defense, and currently serves as a member of the MacArthur Foundation Research Network on Mental Health and the Law. He received a J.D. degree in 1964 from the New York University School of Law.

Judith J. Wurtman is a research scientist at the Massachusetts Institute of Technology. She completed her undergraduate studies at Wellesley College and her Ph.D. in cell biology at George Washington University. Dr. Wurtman is the author of numerous books and articles, including *The Carbohydrate Craver's Diet* and *Managing Your Mind and Mood with Food*. With Richard Wurtman, she coedits the series *Nutrition and the Brain*.

Richard J. Wurtman is Professor of Neuroscience and Director of the Massachusetts Institute of Technology's Clinical Research Center. After completing undergraduate work at the University of Pennsylvania, he graduated from the Harvard Medical School. He did clinical training in medicine at Massachusetts General Hospital and research training in neurochemistry at the National Institute of Mental Health prior to coming to MIT 25 years ago. Professor Wurtman is the author of numerous scientific papers and, with Judith Wurtman, coeditor of the series *Nutrition and the Brain*.

Arthur Yuwiler received his Ph.D. in chemistry from the University of California at Los Angeles under M. S. Dunn. After three years at the University of Michigan with R. W. Gerard, he returned to California where he is a Veterans Administration research scientist, chief of neurobiochemistry research at the Veterans Administration Medical Center, West Los Angeles, and Professor of Psychiatry and Biobehavioral Sciences at UCLA.

K. C. Yuwiler received his J.D. from Southwestern University and was admitted to the bar in California, and for the U.S. District Court, Eastern District of California, in 1986 and for the U.S. District Court, Central and North Districts of California, and U.S. Court of Appeals, Ninth Circuit, in 1987. He is in practice as an associate at Rick Edwards, Inc. Century Park East, Los Angeles.

Index

Acetylcholine (ACh), 165

ADD. *See* Attention deficit disorder

Aggression (*see also* Behavior): association of drugs with, 138–41; association of serotonin levels with, 29–30; and dominance compared, 134; effects of dopamine on, 120–21, 124–25; factors associated with, 115–16, 142–43; relationship between CSF 5-HIAA levels and, 135–38, 141–42

Agren, H., 86

AIDS, 217

Alcohol (*see also* Alcoholism): possible impact of, on the brain, 172

Alcoholism: among arsonists, 88; and behavior, 231, 232; and impulsive violent behavior, 80–81; type-2, and recidivism, 27

Alexander, Richard, 8

ALI. *See* American Law Institute

American Law Institute (ALI), 171, 216–17, 218–19

Amphetamine, 121

Analysis of covariance (ANCOVA), 72

Analysis of variance (ANOVA), 72

ANCOVA. *See* Analysis of covariance

ANOVA. *See* Analysis of variance

Aristotle, 5, 6

Arsonists: association of CSF monoamine metabolite levels with, 70, 71–75, 77–78

Asberg, M., 50–51, 52, 53, 62, 78, 86

Attention deficit disorder (ADD), 234–35, 236

Axelrod, Julius, 103

Ballenger, J. C., 51, 52, 62, 86

Bartko, J., 220n.2

Baselon, David, 169

Bear, David, 172–73

Bea v. Greaves, 240

Beccaria, C. B., 161

Behavior (*see also* Aggression; Compulsivity; Dominance; Impulsivity; *and specific behaviors*), 5, 6, 19; association of CSF 5-HIAA levels with violent types of, 62–67; association of neurotransmitter activity with criminal types of, 27; causation and criminal types of, 30, 197–98n.30; changing norms of, 113–14; control of, 187, 199n.45; implications of effect of serotonin on, 231–33; new sciences of, 163–65; scientific model of, 162

Bentham, Jeremy, 161

Berkelmans, B., 119

Bertilsson, L., 62, 78

Bester, Alfred, 191

Betrayal, 208

Bevan, P., 119

Biology, 4; impact of evidence from, on the law, 20–22, 31–34, 45–46, 54–56, 125–26, 229–30; importance of, in understanding human behavior and